Antibacterial Surfaces

Elena P. Ivanova • Russell J. Crawford
Editors

Antibacterial Surfaces

Springer

Editors
Elena P. Ivanova
School of Science
Faculty of Science, Engineering
and Technology
Swinburne University of Technology
Hawthorn, Australia

Russell J. Crawford
School of Science
Faculty of Science, Engineering
and Technology
Swinburne University of Technology
Hawthorn, Australia

ISBN 978-3-319-18593-4 ISBN 978-3-319-18594-1 (eBook)
DOI 10.1007/978-3-319-18594-1

Library of Congress Control Number: 2015956025

Springer Cham Heidelberg New York Dordrecht London

Printed on acid-free paper

Springer International Publishing AG Switzerland is part of Springer Science+Business Media (www.springer.com)

Contents

Chapter 1
Introduction to Antibacterial Surfaces

Hayden K. Webb, Russell J. Crawford, and Elena P. Ivanova

Abstract Colonisation of surfaces by bacteria occurs commonly in the environment. When this colonisation occurs on materials that are used in modern civilization, it is most often a detrimental process. This can range from bacteria being responsible for the infection of medical implants or tissues causing infection to humans or animals, to biological layers being built-up on ships or in air-conditioning systems causing increased drag and fuel costs. In each case, the result is undesirable, and therefore it is highly desirable to identify ways by which the growth of bacteria on surfaces can be eliminated or controlled. This can be achieved through preventing the initial adhesion of cells, and/or by killing any cells that are able to attach to the surface. In this chapter, a brief overview is provided regarding some of the issues associated with the attachment of bacteria to surfaces, together with a description of the main strategies currently being employed for controlling the initial attachment processes. These strategies will be expanded upon in the subsequent chapters.

Keywords Antibacterial surfaces • Antibiofouling • Mechanobactericidal surfaces • Superhydrophobicity • Antibiotic resistance • Biofilm formation

1.1 The Definition of Antibacterial Surfaces

As a concept, the term ‘antibacterial surface’ is generally well understood. There are, however, one or two subtle clarifications to this term that should be addressed, for the sake of properly defining the term for its use in the subsequent chapters. The term ‘antibacterial surface’ applies to the surface of any material or agent that works to prevent or limit the growth and proliferation of bacteria (Hasan et al. 2013). This is an important distinction, as the word ‘antibacterial’ is commonly understood as being the effective of killing bacterial cells. Bactericidal surfaces certainly fall

---

H.K. Webb • R.J. Crawford • E.P. Ivanova (✉)
School of Science, Faculty of Science, Engineering and Technology,
Swinburne University of Technology, P.O. Box 218, Hawthorn 3122, Victoria, Australia
e-mail: eivanova@swin.edu.au

E.P. Ivanova, R.J. Crawford (eds.), *Antibacterial Surfaces*,
DOI 10.1007/978-3-319-18594-1_1

under the umbrella of being 'antibacterial', however 'bacteriostatic' and 'antibiofouling' surfaces can also be regarded as 'antibacterial' in their action. Each of these concepts will be addressed in greater detail in the following sections.

1.2 The Ancient Battle Against Bacteria

Bacteria are ubiquitous, being found in most places on the Earth and therefore it is necessary for all other organisms to develop strategies to coexist with them. These strategies vary greatly, from some organisms having developed self-cleaning surfaces that limit their need to interact with bacteria, the development of chemical defence mechanisms such as an immune system, and developing the ability to utilise bacteria as a food source. Mankind is not exempt from this requirement. Throughout history, humans have battled against the detrimental effects associated with bacteria, some of which have been, and are still responsible for some of the most deadly diseases known. The Bubonic Plague, also known as 'Black Death', resulted from humans being infected by *Yersinia pestis*, a bacterium from the *Enterobacteriaceae* family. This plague is famous for killing a third of the population of Europe (approximately 25 million people) in the fourteenth century (Haensch et al. 2010). It was suggested that the Austrian composer Mozart died as a result of a streptococcal infection at the age of 35 (Zegers et al. 2009). Of course, bacterial diseases continue to be a major concern in modern times. *Mycobacterium tuberculosis*, the bacterium responsible for tuberculosis, infected 8.6 million people in 2012, leading to 1.3 million deaths (World Health Organization 2015). In addition, post-operative infections caused by bacteria such as methicillin-resistant *Staphylococcus aureus* (MRSA) and *Pseudomonas aeruginosa* are a leading cause of death from complications arising after surgery (Boles and Horswill 2011; Lowy 2003; Klevens et al. 2007).

Pathogenesis and disease, while certainly being extremely important, are not the only undesirable bacteria-associated problems. Food spoilage by bacteria is commonplace, and many food preservation techniques such as salting and pasteurisation were developed in an effort to preserve food for greater periods of time. Adhesion of bacteria and their subsequent multiplication on the surfaces of substrate materials can lead to a wide range of problems, usually in the form of added mass, blockage or in the reduction of aero- or hydrodynamic properties. For example, the shipping industry incurs significant expenses in cleaning ships where bacteria and other organisms have attached to their hulls, causing decreased fuel efficiency (Schultz et al. 2011). Bacterial growth in water-cooled air conditioning systems can result in a loss of heat transfer efficiency and therefore results in an overall decreased effectiveness of the device (Ager and Tickner 1983; Gilbert et al. 2010). Many bacteria have also now been implicated in the phenomenon of biocorrosion, whereby cells that have attached to metallic surfaces (e.g. copper water pipes) lead to the corrosion of the material through the formation of 'pits' in the surface (Beech and Sunner 2004; Gu et al. 2009a, b). The problem of how to control bacterial adhesion and growth is one that has been posed for a long time, and has no simple solution has, as yet, been identified for a number of reasons, discussed as follows:

1.2.1 Extreme Adaptability of Bacteria

The first and most important factor that makes bacteria difficult to control is their high metabolic and physiological adaptability (Bos et al. 1999; Callow and Callow 2002). Bacteria have a high propensity for acquiring new genetic material. They generally can double their population in a short time, with each generation having the potential to develop new mutations, which may be beneficial, detrimental or neutral to the further proliferation of the bacteria. Horizontal gene transfer is also commonplace among bacteria, whereby genetic material is exchanged in the form of plasmids via pili linkages (Dalton and March 1998; Kjelleberg and Molin 2002; Daniels et al. 2004). Some bacteria are capable of taking up extracellular DNA and incorporating it into their transcriptomes and some viruses have the ability to introduce new genetic elements into a bacterial cell via transfection (Thomas and Nielsen 2005; Ochman et al. 2000). Together, these factors enable bacteria to tune their genetic makeup, and through natural selection they can exploit changes in their environment to establish an ecological niche. This has led to the ability of bacteria to colonise a wide variety of environments, including the harsh conditions present in deep sea vents, and their ability to utilise numerous sources of carbon (Baum et al. 2002; Hay 1996).

1.2.2 Antibiotic Resistance

Antibiotic resistance is a growing public health concern, which has arisen as a direct result of the high adaptability of bacteria. Through one of the mechanism listed in Sect. 1.2.1, bacterial cells may acquire the necessary properties to escape the actions of one or more antibiotics. Administration of antibiotics then eliminates the competition to the antibiotic resistant cells, which can then flourish (Fig. 1.1). Naturally, further doses of the same antibiotic no longer have an equivalent effect (Gilbert et al. 2010; Lowy 2003; Simões et al. 2009).

The World Health Organization considers antimicrobial resistance to be one of the three greatest threats to human health (World Health Organization 2014). In addition to this increased resistance, causing a decreased ability to treat established infections, rising antimicrobial resistance has flow-on effects to other treatments that rely on antibiotics to prevent infections. For example, any immunosuppressive therapies, such as chemotherapy, and implantation of biomedical devices become considerably more dangerous without antibiotics.

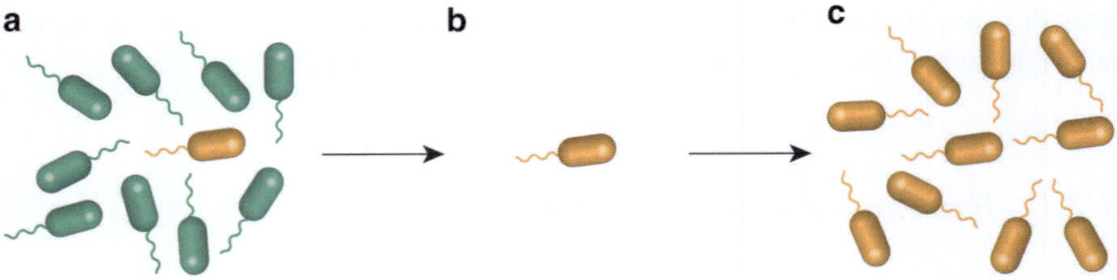

Fig. 1.1 (**a**) Within the natural microbiome, any given bacterium may acquire a gene resistance to a given antibiotic. (**b**) On administration of the antibiotic, all of the sensitive cells are eliminated, removing any nutritional competition for the resistant cell. (**c**) The resistant cell can then flourish, and the microbial community regenerates itself, this time with widespread antibiotic resistance

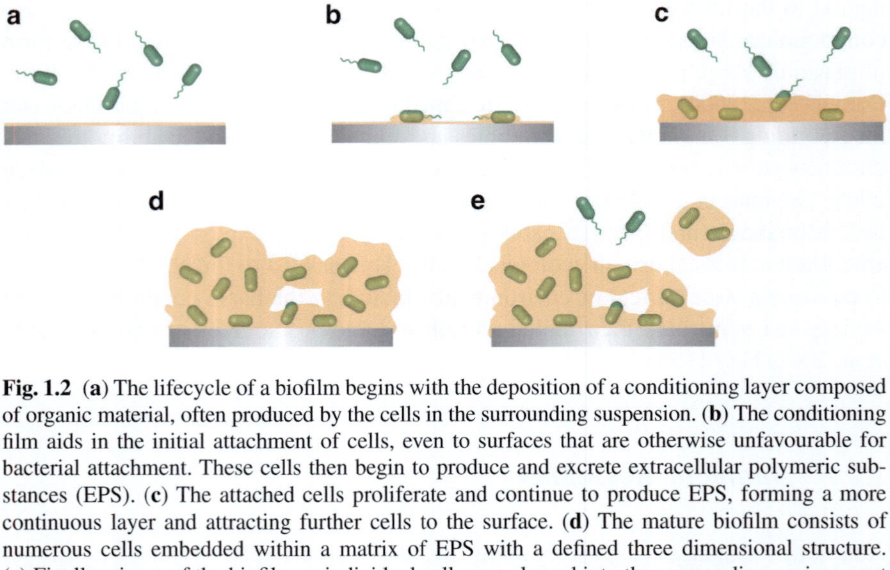

Fig. 1.2 (**a**) The lifecycle of a biofilm begins with the deposition of a conditioning layer composed of organic material, often produced by the cells in the surrounding suspension. (**b**) The conditioning film aids in the initial attachment of cells, even to surfaces that are otherwise unfavourable for bacterial attachment. These cells then begin to produce and excrete extracellular polymeric substances (EPS). (**c**) The attached cells proliferate and continue to produce EPS, forming a more continuous layer and attracting further cells to the surface. (**d**) The mature biofilm consists of numerous cells embedded within a matrix of EPS with a defined three dimensional structure. (**e**) Finally, pieces of the biofilm or individual cells are released into the surrounding environment and are able to colonise new areas

1.2.3 Biofilm Formation

Biofilms are three-dimensional communities of bacteria embedded within an extracellular matrix that is usually composed primarily of polysaccharides. Formation of biofilms is an old and very effective method by which bacteria grow and proliferate. The initial formation begins with the colonisation of a surface by individual cells, which may be the same or different species. The cells that initially adhere to the surface start to grow and divide, and once the cell density reaches a certain level, the cells secrete extracellular polymeric substances (EPS) that form the biofilm matrix. As the biofilm matures, it develops channels to allow the intake of nutrients and the expulsion of wastes. At the final stage of the biofilm cycle, small pieces or individual cells either break off or are released from the biofilm, and may travel to and colonise new regions (Fig. 1.2). Biofilms provide exceptional mechanical and chemical resistance to the cells that they contain, and therefore once biofilms have been formed, they are exceedingly difficult to remove (O'Toole et al. 2000; Schmidt et al. 2012).

1.3 Strategies for Coping with Bacteria

Much progress has been made in preventing or mitigating the detrimental impact of bacterial contamination over a long period of time. The ancient Egyptians, Greeks and Aztecs all used copper-containing compounds and alloys as methods for sterilising wounds and drinking water (Michels et al. 2005). Silver has also proven to be highly effective at killing bacterial cells (Ivanova et al. 2011; Rai et al. 2009). The discovery of penicillin, in particular, was a huge step forward in fighting bacterial infections (Chain et al. 1940). This opened the door for the production of several generations of antibiotic compounds, subsequently revolutionising medicine.

The production of surfaces that are specifically antibacterial in their action is a somewhat more contemporary approach, and in recent times the field has enjoyed increased interest from researchers, with numerous advances made in this technology. With the recognition of the fact that once established, biofilms are extremely difficult to eradicate came the trend of developing methods for the prevention of bacterial attachment, rather than identifying ways to remedy surfaces that contained attached biofilms. Inspired by the work of Bartholott and Neinhuis in 1997 on the lotus leaf, many researchers have focussed on the investigation and production of surfaces that are able to remain clean in a similar way to the lotus (Barthlott and Neinhuis 1997). Such self-cleaning surfaces typically have special properties with regard to their wettability, i.e. superhydrophobicity, which aids in the removal of contaminants. Several other natural surfaces that possess similar properties have since been discovered (Gao and Jiang 2004; Watson et al. 2011; Bixler et al. 2014; Sun et al. 2009), and many more have been artificially fabricated (Zhang et al. 2006; Lee et al. 2004; Guo et al. 2012). Surfaces such as these, that prevent the adhesion and attachment of microbes, are termed 'antibiofouling surfaces', or often simply 'antifouling surfaces'.

Prior to the research being carried out on the development of antibiofouling surfaces, the common approach for developing antibacterial surfaces was to evoke bactericidal effects, i.e. to modify or functionalise the surface so that it had the capability to quickly kill bacterial cells that came into contact or close proximity to the surface, preventing them from proliferating and initiating the formation of a biofilm. Coating or functionalising surfaces with compounds known to be bactericidal, e.g. silver or antibiotics (Gazit 2007; Qi et al. 2004; In et al. 2007; Maness et al. 1999), was an early effective approach, however leaching and releasing bioactive components often occurs, which may present biosafety challenges and reduction in the longevity of the material. Polycationic polymers have also been shown to be effective bactericidal materials that possess some degree of durability (Haldar et al. 2008; Yang et al. 2014). Bactericidal surfaces have again come into focus with the discovery of a class of materials that have a novel mechanism for killing cells. The wing of the *Psaltoda claripennis* cicada was the first material reported to kill bacterial cells as a result of the physical nanostructure of the surface, with no apparent influence arising from the chemistry of the surface (Hasan et al. 2012; Ivanova et al. 2012; Pogodin et al. 2013). This mechanism by which this bactericidal activity was achieved was referred to as a 'mechanobactericidal' effect. Other insect wings have since been determined to be bactericidal through using a similar mechanism, and a bioinspired silicon analogue has also been reported (Ivanova et al. 2013).

1.4 Summary

The bacterial colonisation of surfaces can lead to many detrimental effects on human health, industry and the environment. Therefore, a technological need exists for the development of materials whose surfaces resist the build-up of bacteria. A range of approaches are available for the design of such surfaces, and in the following chapters a wide variety of antibacterial surfaces will be described. The prevailing mechanisms that determine their effectiveness, methods of fabrication in the case of synthetic surfaces, and an analysis of the applications in which these types of surfaces are likely to find use, will be discussed.

References

Ager BP, Tickner JA (1983) The control of microbiological hazards associated with air-conditioning and ventilation systems. Ann Occup Hyg 27(4):341–358

Barthlott W, Neinhuis C (1997) Purity of the sacred lotus, or escape from contamination in biological surfaces. Planta 202(1):1–8

Baum C, Meyer W, Stelzer R, Fleischer LG, Siebers D (2002) Average nanorough skin surface of the pilot whale (*Globicephala melas*, Delphinidae): considerations on the self-cleaning abilities based on nanoroughness. Mar Biol 140(3):653–657

Beech IB, Sunner J (2004) Biocorrosion: towards understanding interactions between biofilms and metals. Curr Opin Biotechnol 15(3):181–186

Bixler GD, Theiss A, Bhushan B, Lee SC (2014) Anti-fouling properties of microstructured surfaces bio-inspired by rice leaves and butterfly wings. J Colloid Interface Sci 419:114–133

Boles BR, Horswill AR (2011) Staphylococcal biofilm disassembly. Trends Microbiol 19(9):449–455

Bos R, Van Der Mei HC, Busscher HJ (1999) Physico-chemistry of initial microbial adhesive interactions – its mechanisms and methods for study. FEMS Microbiol Rev 23(2):179–229

Callow ME, Callow JA (2002) Marine biofouling: a sticky problem. Biologist 49(1):10–14

Chain E, Florey HW, Adelaide MB, Gardner AD, Oxfd DM, Heatley NG, Jennings MA, Orr-Ewing J, Sanders AG (1940) Penicillin as a chemotherapeutic agent. Lancet 236(6104):226–228

Dalton HM, March PE (1998) Molecular genetics of bacterial attachment and biofouling. Curr Opin Biotechnol 9(3):252–255

Daniels R, Vanderleyden J, Michiels J (2004) Quorum sensing and swarming migration in bacteria. FEMS Microbiol Rev 28(3):261–289

Gao X, Jiang L (2004) Water-repellent legs of water striders. Nature 432(7013):36

Gazit E (2007) Self-assembled peptide nanostructures: the design of molecular building blocks and their technological utilization. Chem Soc Rev 36(8):1263–1269

Gilbert Y, Veillette M, Duchaine C (2010) Airborne bacteria and antibiotic resistance genes in hospital rooms. Aerobiologia 26(3):185–194

Gu X, Zheng Y, Cheng Y, Zhong S, Xi T (2009a) In vitro corrosion and biocompatibility of binary magnesium alloys. Biomaterials 30(4):484–498

Gu XN, Zheng W, Cheng Y, Zheng YF (2009b) A study on alkaline heat treated Mg-Ca alloy for the control of the biocorrosion rate. Acta Biomater 5(7):2790–2799

Guo F, Su X, Hou G, Liu Z, Mei Z (2012) Fabrication of superhydrophobic TiO_2 surface with cactus-like structure by a facile hydrothermal approach. Colloids Surf A Physicochem Eng Asp 395:70–74

Haensch S, Bianucci R, Signoli M, Rajerison M, Schultz M, Kacki S, Vermunt M, Weston DA, Hurst D, Achtman M, Carniel E, Bramanti B (2010) Distinct clones of *Yersinia pestis* caused the Black Death. PLoS Pathog 6:e1001134

Haldar J, Chen J, Tumpey TM, Gubareva LV, Klibanov AM (2008) Hydrophobic polycationic coatings inactivate wild-type and zanamivir- and/or oseltamivir-resistant human and avian influenza viruses. Biotechnol Lett 30(3):475–479

Hasan J, Webb HK, Truong VK, Pogodin S, Baulin VA, Watson GS, Watson JA, Crawford RJ, Ivanova EP (2012) Selective bactericidal activity of nanopatterned superhydrophobic cicada *Psaltoda claripennis* wing surfaces. Appl Microbiol Biotechnol 97(20):9257–9262

Hasan J, Crawford RJ, Ivanova EP (2013) Antibacterial surfaces: the quest for a new generation of biomaterials. Trends Biotechnol 31:295–304

Hay ME (1996) Marine chemical ecology: what's known and what's next? J Exp Mar Biol Ecol 200(1–2):103–134

In S, Orlov A, Berg R, García F, Pedrosa-Jimenez S, Tikhov MS, Wright DS, Lambert RM (2007) Effective visible light-activated B-doped and B, N-codoped TiO_2 photocatalysts. J Am Chem Soc 129(45):13790–13791

Ivanova EP, Hasan J, Truong VK, Wang JY, Raveggi M, Fluke C, Crawford RJ (2011) The influence of nanoscopically thin silver films on bacterial viability and attachment. Appl Microbiol Biotechnol 91:1149–1157

Ivanova EP, Hasan J, Webb HK, Truong VK, Watson GS, Watson JA, Baulin VA, Pogodin S, Wang JY, Tobin MJ, Löbbe C, Crawford RJ (2012) Natural bactericidal surfaces: mechanical rupture of *Pseudomonas aeruginosa* cells by cicada wings. Small 8(16):2489–2494

Ivanova EP, Hasan J, Webb KW, Gervinskas G, Juodkazis S, Truong VK, Wu AHF, Lamb RN, Baulin VA, Watson GS, Watson JA, Mainwaring DE, Crawford RJ (2013) Bactericidal activity of black silicon. Nat Commun 4:2838–2844

Kjelleberg S, Molin S (2002) Is there a role for quorum sensing signals in bacterial biofilms? Curr Opin Microbiol 5(3):254–258

Klevens RM, Morrison MA, Nadle J, Petit S, Gershman K, Ray S, Harrison LH, Lynfield R, Dumyati G, Townes JM, Craig AS, Zell ER, Fosheim GE, McDougal LK, Carey RB, Fridkin SK (2007) Invasive methicillin-resistant *Staphylococcus aureus* infections in the United States. J Am Med Assoc 298(15):1763–1771

Lee W, Jin MK, Yoo WC, Lee JK (2004) Nanostructuring of a polymeric substrate with well-defined nanometer-scale topography and tailored surface wettability. Langmuir 20(18):7665–7669

Lowy FD (2003) Antimicrobial resistance: the example of *Staphylococcus aureus*. J Clin Investig 111(9):1265–1273

Maness PC, Smolinski S, Blake DM, Huang Z, Wolfrum EJ, Jacoby WA (1999) Bactericidal activity of photocatalytic TiO_2 reaction: toward an understanding of its killing mechanism. Appl Environ Microbiol 65(9):4094–4098

Michels HT, Wilks SA, Noyce JO, Keevil CW (2005) Copper alloys for human infectious disease control. In: Proceedings of the materials science and technology conference, copper for the 21st century symposium, Pittsburgh, PA, September, pp 1–13
Ochman H, Lawrence JG, Grolsman EA (2000) Lateral gene transfer and the nature of bacterial innovation. Nature 405(6784):299–304
O'Toole G, Kaplan HB, Kolter R (2000) Biofilm formation as microbial development. Annu Rev Microbiol 54:49–79
Pogodin S, Hasan J, Baulin VA, Webb HK, Truong VK, Nguyen THP, Boshkovikj V, Fluke CJ, Watson GS, Watson JA, Crawford RJ, Ivanova EP (2013) Biophysical model of bacterial cell interactions with nanopatterned cicada wing surfaces. Biophys J 104(4):835–840
Qi L, Xu Z, Jiang X, Hu C, Zou X (2004) Preparation and antibacterial activity of chitosan nanoparticles. Carbohydr Res 339(16):2693–2700
Rai M, Yadav A, Gade A (2009) Silver nanoparticles as a new generation of antimicrobials. Biotechnol Adv 27(1):76–83
Schmidt S, Winter J, Gallert C (2012) Long-term effects of antibiotics on the elimination of chemical oxygen demand, nitrification, and viable bacteria in laboratory-scale wastewater treatment plants. Arch Environ Contam Toxicol 63(3):354–364
Schultz MP, Bendick JA, Holm ER, Hertel WM (2011) Economic impact of biofouling on a naval surface ship. Biofouling 27(1):87–98
Simões M, Simões LC, Vieira MJ (2009) Species association increases biofilm resistance to chemical and mechanical treatments. Water Res 43(1):229–237
Sun M, Watson GS, Zheng Y, Watson JA, Liang A (2009) Wetting properties on nanostructured surfaces of cicada wings. J Exp Biol 212(19):3148–3155
Thomas CM, Nielsen KM (2005) Mechanisms of, and barriers to, horizontal gene transfer between bacteria. Nat Rev Microbiol 3(9):711–721
Watson GS, Cribb BW, Watson JA (2011) Contrasting micro/nano architecture on termite wings: two divergent strategies for optimising success of colonisation flights. PLoS ONE 6(9):Art. No. e24368:1–10
World Health Organization (2015) Tuberculosis. Fact sheet No 104. Retrieved 20 Aug 2015
World Health Organization (2014) Antimicrobial resistance. Global report on surveillance. Retrieved 27 Oct 2014
Yang WJ, Neoh KG, Kang ET, Teo SLM, Rittschof D (2014) Polymer brush coatings for combating marine biofouling. Prog Polym Sci 39(5):1017–1042
Zegers RHC, Weigl A, Steptoe A (2009) The death of Wolfgang Amadeus Mozart: an epidemiological perspective. Ann Intern Med 151:274–278
Zhang G, Zhang J, Xie G, Liu Z, Shao H (2006) Cicada wings: a stamp from nature for nanoimprint lithography. Small 2(12):1440–1443

Chapter 2
Natural Antibacterial Surfaces

Song Ha Nguyen, Hayden K. Webb, Russell J. Crawford, and Elena P. Ivanova

Abstract The world has long experienced the impact of surfaces fouling with biofilms, not only in economic terms, but also, importantly, the adverse effect that biofilms can have with regard to public health. In the USA alone, billions of dollars are spent every year cleaning equipment, decontaminating products and cleaning ship hulls, while over 100,000 mortalities are reported annually as a result of infections resulting from medical device implant surgeries that have been compromised by the presence of pathogenic bacteria. Of great concern is that the heavy use of chemicals for neutralising bacterial colonies has resulted in the production of tougher, more resistant strains of pathogenic bacteria, which challenges the scientific community to find new approaches for controlling the formation of biofilms. Recently, the hierarchical structures found on the surfaces of some organisms, such as plant leaves and insect cuticles, have been shown to be superhydrophobic, self-cleaning, and possess bactericidal activity. Since the self-cleaning properties of the lotus leaf were reported in 1997, there has been a great deal of effort put into exploring this approach as a potential method for controlling the formation of biofilms. These discoveries may provide alternative approaches for controlling bacterial behaviour, either before or after the bacteria have attached to a substrate surface. This chapter provides a summary of some of the strategies employed by nature for controlling the colonisation of bacteria on surfaces.

Keywords Antibiofouling • Superhydrophobicity • Self-cleaning • Bactericidal activity • Wettability • Plant leaves • Insect cuticule • Mechanobactericidal activity

2.1 Introduction

Biofouling has remained a complex, problematic issue for a long period of time. Its consequences impact not only upon the economy, but also public health. For this reason, antibacterial materials have been developed in order to design advanced strategies for limiting the colonisation of bacteria on their surfaces (Zhang et al.

S.H. Nguyen • H.K. Webb • R.J. Crawford • E.P. Ivanova (✉)
School of Science, Faculty of Science, Engineering and Technology,
Swinburne University of Technology, P.O. Box 218, Hawthorn 3122, Victoria, Australia
e-mail: eivanova@swin.edu.au

E.P. Ivanova, R.J. Crawford (eds.), *Antibacterial Surfaces*,
DOI 10.1007/978-3-319-18594-1_2

2013). Traditionally, antibacterial surfaces were designed so that their surfaces would leach biocides, which would kill bacterial cells in situ and in areas surrounding the surface. For example, peptides and chitosan have been used as chemical-based methods for controlling the colonisation of bacteria on surfaces (Gazit 2007; Qi et al. 2004). Antibacterial metal nanoparticles such as silver (Rai et al. 2009), copper (Hsiao et al. 2006), and molybdenum (Yasuyuki et al. 2010) have also been used as an additive for controlling bacterial attachment. The effects of these metals on human health and the environment are of growing concern. In addition, ever-increasing doses are now being required for chemical-based methods to effectively sterilise surfaces. This increased use of chemical agents has led to another problem; bacterial resistance to antibiotics. Therefore, the scientific community must continue to find alternative methods for effectively controlling bacterial attachment.

More recently, new approaches for preventing bacterial attachment, which use photocatalytic metal oxides such as TiO_2 (Gelover et al. 2006) and ZnO (Franklin et al. 2007; Jones et al. 2008), have been developed. These materials produce highly reactive species such as hydroxyl radicals, hydrogen peroxide and superoxide, which are lethal to *Escherichia coli* and some other types of bacterial cells (Maness et al. 1999; Ibáñez et al. 2003). These metal oxides are, however, mainly activated by UVA light sources, which limits their potential biomedical applications (In et al. 2007; Fu et al. 2005).

Superhydrophobic/self-cleaning surfaces based on natural materials such as plant leaves and insect cuticles are currently being developed for controlling bacterial colonisation. Traditionally, only those materials that could induce bacterial cell death were considered to be antibacterial materials (Zhang et al. 2013), however antibiofouling materials, many of which are also superhydrophobic, can also classified in this category due to their potential application in controlling bacterial attachment. Many natural surfaces have been subjected to harsh environmental conditions in that they are constantly in contact with pollutants and changing weather conditions. Over millions of years of evolution, organisms have developed strategies that enable them to survive. Lotus leaves have been studied in detail since 1997 and have given rise to the archetypal "lotus effect" due to their self-cleaning nature (Barthlott and Neinhuis 1997). The properties that afford the lotus leaf these self-cleaning properties are their high water contact angle ($\theta > 150°$) and low tilting angle ($\theta < 10°$), the angle to which the leaf needs to be tilted in order for the water droplet to roll off the surface. These properties allow water droplets to collect dirt as they move over the surface, hence the term 'self-cleaning' (Webb et al. 2011) being applied to such surfaces. If artificial surfaces can be synthetically produced to possess similar surface characteristics and therefore cause water to behave in a similar way, bacterial cells could potentially also be cleaned from such surfaces before they have a chance to develop a biofilm. A similar phenomenon was also observed on the surfaces of sections of some insects, such as cicada and dragonfly wings. Interestingly, some insect surfaces possess not only self-cleaning properties, but also act as bactericidal surfaces (Ivanova et al. 2012; Pogodin et al. 2013).

2.2 Basics of Biofilms

A biofilm is defined as the attachment and development of microorganism community embedded in extracellular matrix on a surface (O'Toole et al. 2000). The organisms undergo a transition state between having the ability to be free swimming in their native environment (planktonic cells) to being cells that form part of the surface-attached community. The essential factors necessary for the formation of a biofilm are microbes and a substratum (Garrett et al. 2008). There are numerous advantages for bacteria to be part of a biofilm; these include resistance to antibiotics (Schmidt et al. 2012) and disinfectants (Ryu and Beuchat 2005; Simões et al. 2009) and being part of a dynamic environment (Liu and Tay 2002; Di Iaconi et al. 2005). Intercellular communication within the biofilm community also enhances the regulation of gene expression, which enables the bacterial cells to temporally adapt to any phenotypic variations in the surrounding environment, in addition to any deficiency in the available nutrient supply (Dalton and March 1998; Kjelleberg and Molin 2002; Daniels et al. 2004).

Biofilm formation can involve a single microbial species or multiple microbial species adhering onto a range of surfaces. On most environmental surfaces, mixtures of various species will dominate the biofilm. It is usually a single bacterial species, however, that is responsible for the infection of medical devices and implants (Holmes et al. 2008; Behlau and Gilmore 2008; Seo et al. 2008; Bulgarelli et al. 2013; Wu et al. 2012). According to a public health report in 2002 (Klevens et al. 2007), approximately 64 % of hospital attending cases resulted from the viable bacterial infection of medical devices and implants. These biofilms have been associated with 100,000 mortalities annually in the US alone. Researchers began studying biofilms over three decades ago, with the discovery that under natural living conditions, microorganisms dominantly attach themselves to surfaces (Geesey et al. 1977). The first recorded observation was published in 1933 by Henrici (1933), however the impact of biofilm formation had been recognised even before this time in the form of the fouling of ships in marine environments (Angst 1923). It has been estimated that the fouling of US Navy ships costs approximately US$ 180M–$260M per year. This represents only 0.5 % of the total number of ships world-wide (Schultz et al. 2011).

The initial development of a biofilm is described by a two stage kinetic binding model (Fig. 2.1). The first stage involves the initial reversible interaction that takes place between bacterial cells and the material surface, followed by the second stage where specific and non-specific interactions take place at the molecular level (Lichter et al. 2009; Bos et al. 1999). The interactions that occur in the second stage involve proteins that are expressed on the bacterial surface and on molecules on the material surfaces. The second stage occurs slowly and is irreversible once a mature biofilm has been formed. Apart from these two main steps of biofilm maturation, O'Toole et al. proposed that the starvation response pathway can also be considered as part of the biofilm development (O'Toole et al. 2000). This pathway is developed when the source of nutrients becomes depleted, and single microbial cells detach from the surface and return to their planktonic state, and commence infecting new areas of the surface.

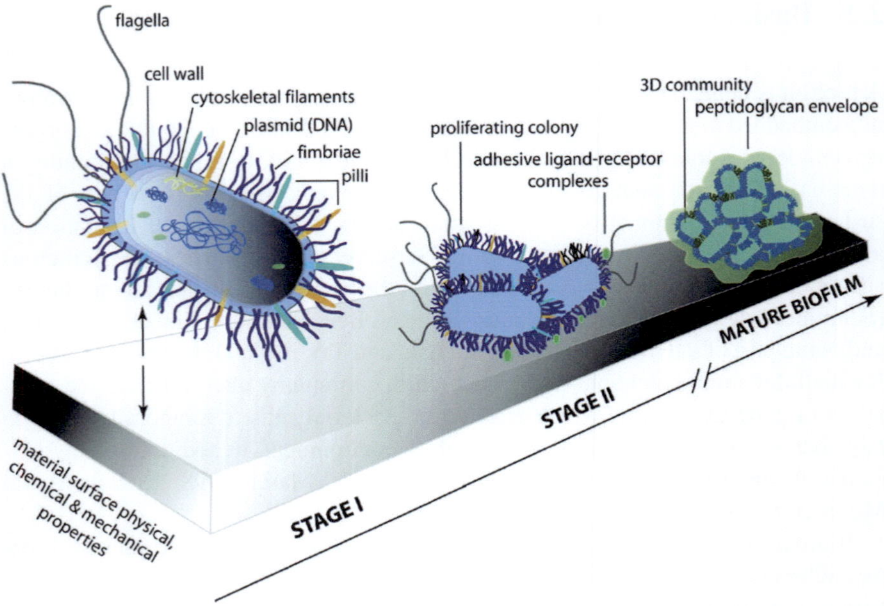

Fig. 2.1 Schematic representation of the two stages of biofilm formation on substrate surfaces (Reproduced with permission from American Chemical Society (Lichter et al. 2009))

For these reasons, controlling bacterial attachment on material surfaces has been a long-standing battle for science. Several approaches have been developed to limit the colonisation of microbes onto the surfaces, however most of these have focussed on using chemical-based methods for bacterial control, which has led to the new and rising problem of bacterial resistance to these agents. Preventing bacterial adhesion from occurring by modifying the surface topography of substrates has been identified as an approach that may provide attractive alternative strategies for controlling biofilm formation.

2.3 Antibiofouling Based on the Self-Cleaning Properties of a Surface

2.3.1 Wettability Theory

Wettability is a measure of the ability of a liquid to wet and spread over a solid surface. The contact angle, which is a function of surface energy of the solid, is formed when the liquid/vapour interface meets that of the liquid/solid interface. The wettability of solid surfaces plays an important role in daily life, industry and agriculture. Surfaces with special wettability properties, exhibiting for example high degrees of hydrophobicity or hydrophilicity, have been the subject of much research due to the potential advantages associated with these types of surfaces (Nosonovsky and Bhushan 2005, 2007; Su et al. 2010). According to the most commonly agreed definitions, surfaces can be classified into one of four categories:

- surfaces with a water contact angle greater than 150° and a tilting angle less than 10°. These surfaces are considered to be superhydrophobic and self-cleaning.
- surfaces with a water contact angle between 90° and 150°. These surfaces are described as being hydrophobic.
- surfaces with a water contact angle between 10° and 90°. These surfaces are described as being hydrophilic.
- surfaces with a water contact angle less than 10°. These surfaces are considered to be superhydrophilic.

The measurement of water contact angle (WCA, θ) is the most common method for determining the wettability of surfaces. Originally, the contact angles were determined by Young's Eq. 2.1 where the surfaces were assumed to be smooth, rigid, chemically homogeneous, insoluble and non-reactive (Zhang et al. 2013; Young 1805):

$$\cos\theta = \frac{\gamma_{sv} - \gamma_{sl}}{\gamma_{lv}} \tag{2.1}$$

where θ is the contact angle; γ is the surface tension which is determined as the force per unit length; s, l, v represent solid, liquid and vapour, respectively. Surface tension is also known as surface energy, which is the energy required to break an intermolecular bond (Nosonovsky and Bhushan 2008). Numerically, surface tension and surface energy are equivalent, however they are thermodynamically different (Yan et al. 2011). Surface tension is used when dealing with liquids, whilst surface energy is a general term used for the description of solid surfaces.

In practice, most surfaces are both rough and chemically heterogeneous, and this complexity at the interface between the solid and liquid surfaces causes difficulties in determining the real contact angle. Wenzel first proposed a model to explain the relationship between surface roughness and the measured contact angle (Wenzel 1949), while Cassie and Baxter (1944) described the relationship between the surface fractions of different chemical composition and the contact angle. Wenzel's equation is shown as:

$$\cos\theta_{rough} = r\cos\theta_{smooth} \tag{2.2}$$

where θ_{rough} and θ_{smooth} are water contact angles on rough and ideal smooth surfaces, respectively, and r is the Wenzel roughness factor. The roughness factor is calculated as the ratio between the actual surface area and the projected surface area, which can be used to explain the change in surface hydrophobicity that arises through roughness, not surface chemistry. According to the theory, there are two separate cases where θ_{rough} will behave differently as the roughness factor increases, depending on the value of θ_{smooth}:

(i) if θ_{smooth} <90°, as r increases, θ_{rough} will reduce to 0°
(ii) if θ_{smooth} >90°, as r increases, θ_{rough} will approach 180°.

According to the Cassie and Baxter model, surface heterogeneity induces air entrapment between the topographical structures on a surface, which causes increased surface hydrophobicity, as given in the equation:

$$\cos\theta = f_1 \cos\theta_1 + f_2 \cos\theta_2 \tag{2.3}$$

where, θ is the composite contact angle of the heterogeneous surface, f_1 and f_2 are the area fractions of surface components 1 and 2, and θ_1 and θ_2 are their respective contact angles. This equation has been used widely to explain and/or predict the hydrophobicity of the surfaces with both a micro- and nano-hierarchical structure. When a water droplet sits on a rough surface, the two surface components that affect surface wettability are the surface itself and the air trapped between the surface features. Since the water contact angle on air can be taken as 180° (i.e. $\theta_2 = 180°$), and $f_2 = 1 - f_1$, then Eq. 2.3 becomes:

$$\cos\theta = f_1(\cos\theta_1 + 1) - 1 \tag{2.4}$$

According to Cassie-Baxter theory, superhydrophobicity arises from the combination of hierarchical surface structures that enable the entrapment of air on low surface energy materials. The sliding angle, another parameter that is important in determining the degree of hydrophobicity, is defined as the critical angle at which the water droplets start to slide along a tilted surface (Bhushan et al. 2009; Jung and Bhushan 2006; Yan et al. 2011). The scientific community has become aware of this principle only in the past century, whereas nature has adapted and evolved over millions of years to develop mechanisms that function according to this principle. Lotus leaves have long been regarded as a symbol of purity in many Asian cultures, and this originates from their clean nature despite being often found in unclean environmental conditions. It is now well established that the self-cleaning ability of the lotus leaf is a direct result of surface micro- and nanostructures that maximise the quantity of entrapped air in the surface, resulting in the condition of superhydrophobicity, in accordance to the Cassie-Baxter wetting regime. Several other organisms have been identified to utilise similar mechanisms, including other plant species and some insects. Some marine organisms are also known to remain clean through the different, but related concept of superoleophobicity. The following sections will focus on these organisms and the mechanisms by which their surfaces exhibit antibiofouling properties for controlling bacterial colonisation onto the surfaces.

2.3.2 Plant Leaves

Since Barthlott and Neinhuis first reported the 'lotus effect', the lotus has become the archetype surface for exhibiting superhydrophobicity and self-cleaning abilities (Barthlott and Neinhuis 1997). Lotus leaves satisfy the two factors that are reflected in the Cassie-Baxter theory. The surface is covered by a layer of lipids, which are low in surface energy. The lipids appear as a layer of multiscale structures that enable a large quantity of air to be trapped in between the surface features. This results in a surface with very high WCA ($\theta \approx 165°$) and low tilting angle, hence the surface can remain clean as the water droplets collect dirt and contaminating particles as they roll off the surface. In Fig. 2.2, a mercury droplet that is spherical in shape can be seen to roll across the surface of a leaf. Contaminants also adhere to the droplet rather than the surface. This demonstrates how superhydrophobic and self-cleaning surfaces can be very useful templates for designing antibiofouling materials.

Many other plants exhibit very similar properties to that of the lotus leaf, for example the Indian canna, taro and cabbage leaves. Plants first moved from water onto land approximately 480–360 million years ago; this was an important event in the history of life as it highlighted the consequences of the evolutionary changes of terrestrial organisms and global environments (Kenrick and Crane 1997). To cope with their new environments, plants developed a protective 'skin', known as the cuticle. The plant cuticle is a thin layer of lipophilic compounds that function as a protective barrier to perform various physiological, ecological and developmental roles. These roles include minimising water loss, reducing the leaching of cellular content, decreasing the adhesion of pathogenic spores and dust, protecting tissues

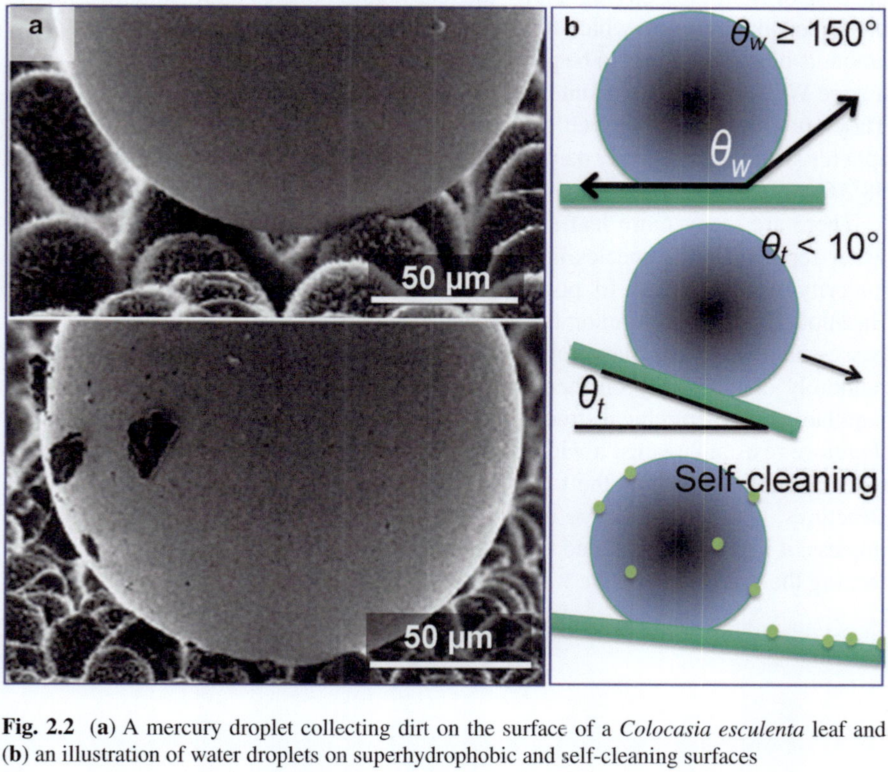

Fig. 2.2 (**a**) A mercury droplet collecting dirt on the surface of a *Colocasia esculenta* leaf and (**b**) an illustration of water droplets on superhydrophobic and self-cleaning surfaces

from ultraviolet radiation, and mediation of their interaction with the surrounding environment (Van Maarseveen et al. 2009). The cuticle contains a continuous extracellular membrane that is made of biopolymers. These polymers cover the primary above-ground organs such as the flowers, leaves, stems and fruit of all land plants (Koch and Ensikat 2008). A mixture of hydrophobic compounds is integrated and superimposed on the cuticles, which is composed of various waxes (Jetter et al. 2000; Barthlott et al. 1998).

Plant waxes that are embedded within the cutin network are called “intracuticular waxes”, whereas “epicuticular waxes” are located on the outer surface of the cuticle (Barthlott and Neinhuis 1997; Barthlott et al. 1998; Buschhaus et al. 2007; Buschhaus and Jetter 2011; Ensikat et al. 2011; Koch et al. 2009). Cutin is a comprised of a polymer of predominantly ω- and mid-chain hydroxyl and epoxy C_{16} and C_{18} fatty acids in addition to glycerol (Samuels et al. 2008). The epicuticular waxes are organised within themselves to form three-dimensional crystals with highly variable morphologies, e.g., nano/micro projections, platelets, rods and tubules (Barthlott et al. 1998; Koch et al. 2006). Some examples of plants with superhydrophobic surfaces are presented in Fig. 2.3.

Both India canna leaves and purple *Setcreasea* are covered by many wax platelets, distributed randomly on a series of rod-like structures. This increases the proportion of air that can be trapped within the surface, producing water contact angles in excess of the 150° contact angle condition for superhydrophobicity (i.e. 165°). In the case of ramee leaves (Fig. 2.3d), the rear face is covered by a randomly distributed fiber-like structure which forms the layers of a web. This also allows for large amounts of entrapped air to be present on the surface, causing the surface to exhibit a large WCA (164°). The front of ramee leaves are significantly different in nature. They are composed of a web of micro-fibers, with many larger micrometer-size spheres without any further nanoscale-structure, and the surface exhibits a WCA of 38° (Guo and Liu 2007).

There are many more leaf surfaces that possess similar properties. Up to 200 water repellent plant species have been screened to measure their WCA and the majority were reported to possess superhydrophobic properties (Neinhuis and Barthlott 1997). The common feature shared by these surfaces is that each of them possessed a very dense layer of three-dimensional cuticular wax crystals arranged randomly or uniformly on their corresponding micro-scale surface features (e.g. papillae). This hierarchical structure enables the plant surfaces to remain clean, and therefore resistant against a wide range of contaminants. Many attempts have been made to understand how the lipids self-assemble into such useful and systematic structures, and while no clear understanding has yet been obtained regarding this process, it has been postulated that the cutin network may act as a template in controlling the orientation of the wax crystals (Jeffree 2006).

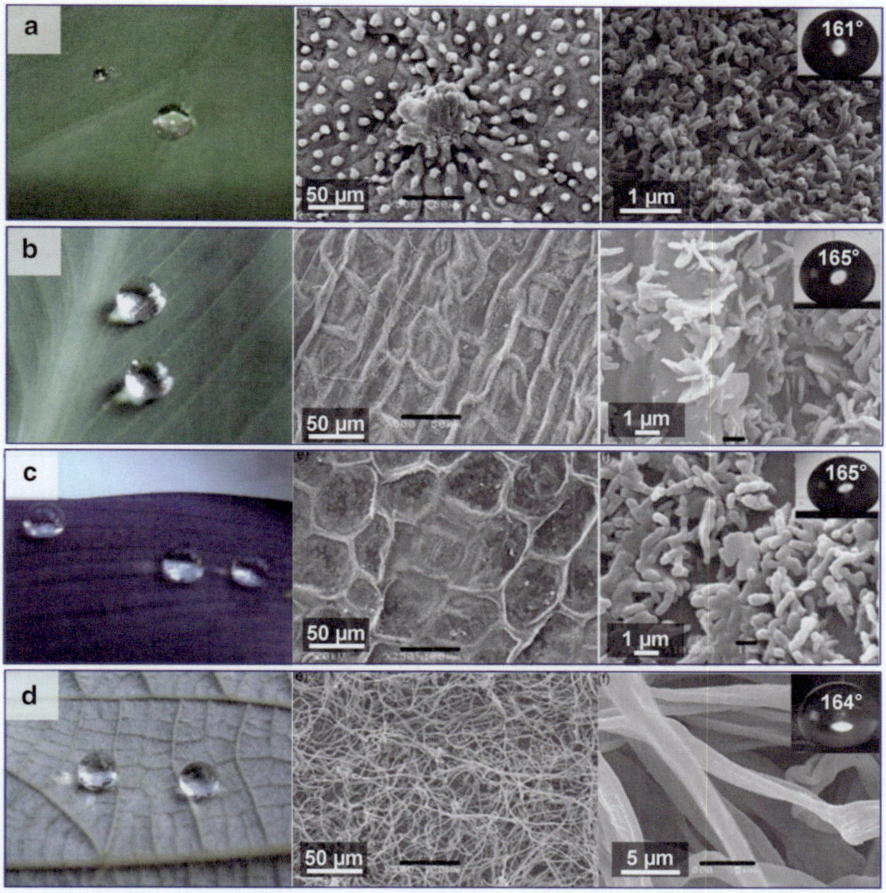

Fig. 2.3 Images of some superhydrophobic plant surfaces, and their corresponding epicuticular wax structures: (**a**) Lotus leaves; (**b**) Indian canna leaves; (**c**) Rear face of purple *Setcreasea* leaves; (**d**) Rear face of ramee leaves (Guo and Liu 2007)

2.3.3 Insect Cuticle

Insects first evolved the ability to fly at least 400 million years ago, and were the first organisms to develop powered flight; taking to the skies at least 90 million years prior to the earliest winged vertebrates (Grimaldi and Engel 2005). Nowadays they represent half of all eukaryotic species on earth. Insect wings are composed of lightweight building materials of thicknesses ranging from 0.5 μm to about 1 mm (Wan et al. 2008; Wootton 1992). In order to adapt to ever-changing environments, insects have evolved to possess geometric, non-smooth structures on their wing surface (Fig. 2.4) (Arsene et al. 2002; Boeve et al. 2011; Nelson and Charlet 2003). The presence of a thin superficial layer of waxes in the epicuticle was first reported by Ramsay in 1935 (Ramsay 1935).

Fig. 2.4 Insect wings and their corresponding surface topographies. (**a**) *Isoptera Nasutitermes* sp.; (**b**) *Hemianax papuensis*; (**c**) *Psaltoda claripennis*; and (**d**) *Lepidoptera papilio xuthus*

As is the case with plant leaves, insect surfaces are covered by a layer of cuticle, which is the barrier that directly interacts with the environment. Their terminology might be different, but in principle they are very similar in construction. The insect cuticle is secreted by a single layer of epidermal cells, forming a lipophilic structure that consists of two major sublayers, which are the epicuticle and the intracuticle (Lockey 1980, 1985; Nelson and Blomquist 1995; Buckner 2010; Jetter and Kunst 2008). The intracuticular layer, located beneath the epicuticle, is a mixture of chitin (poly-N-acetylglucosamine) and protein (Lockey 1980, 1985, 1988). The epicuticle is located in the outermost layer and is composed of a mixture of aliphatic hydrocarbons and their derivatives; these compounds contain one or more oxygenated functional groups including esters, ketones, alcohols, aldehydes and fatty acids (Samuels et al. 2008; Koch and Ensikat 2008). This mixture of organic components is self-organized in the epicuticular layer of the cuticle, a highly-ordered, rough structure, composed of numerous micro- and nanometer-scale features. For some insects, e.g. dragonflies, the epicuticular waxes self-assemble into a three-dimensional layer of "nanopillars", which enable air to be trapped in the spaces between and hence exhibit a high WCA (Ivanova et al. 2013b; Nguyen et al. 2013). Insect wing membranes are composed of lightweight building materials with a thickness ranging from 0.5 µm to approximately 1 mm (Wootton 1992). Their wings are framed by a system of veins that aid in stabilizing the wing as a whole (Kreuz et al. 2001; Gorb 1999; Moussian 2010). The highly-ordered, rough structure of the epicuticle enables insects to minimize their mass but still retain the ability to protect themselves from being wet by rain and coated with pollutants (Fig. 2.4).

A systematic terminology to describe the 2D and 3D micro- and nano-scale structures of the insect cuticle has not thus far been developed. Byun et al. used the terms 'layered cuticle', 'setae', 'denticles' and 'fractal' to describe the morphological features present on the surfaces of the insect wings, and this is the system that will be adopted here. The term 'layered cuticle' refers to a surface that contains scale-like structures that overlap, such as those typically found on butterfly wings. Surfaces with 'setae' contain high aspect ratio nanopillars or hairs. 'Denticle' structures

Table 2.1 Micro- and nano-scale wax crystal morphologies on the epidermal cells of insect wing surfaces and their WCA

Order	Species	Structural morphology	WCA (°)
Isoptera	*Schedorhinotermes* sp.	Setae	71
Coleoptera	*Amphizoa sinica*	Setae	109
Hymenoptera	*Vespa simillima xanthoptera*	Setae	121
Hymenoptera	*Vespa dybowskii*	Setae	126
Hemiptera	*Meimuna microdon*	Denticle	140
Orthoptera	*Atractomorpha lata*	Denticle	148
Orthoptera	*Acrida cinerea cinerea*	Denticle	151
Odonata	*Hemicordulia tau*	Fractal	157
Odonata	*Hemianax papuensis*	Fractal	161
Lepidoptera	*Artogeia canidia*	Layered cuticle	162
Lepidoptera	*Papilio xuthus*	Layered cuticle	168

This table was modified and updated from Byun et al. (2009)

refer to tooth-like projections, and these can vary greatly in their morphology, ranging from small hemispheres to taller nanopillars. 'Fractal' structures are composed of an irregular array of fine nanoscale protrusions (Byun et al. 2009). Among these structural types, the presence of layered cuticles, denticles and fractal structures result in the production of the most hydrophobic surfaces in a majority of cases, whilst the presence of setae alone on a surface usually produces a surface exhibiting hydrophilic properties (Table 2.1).

The superhydrophobicity of an insect wing surface, together with its ability to self-clean, are very important factors that contribute to an insect's ability to survive. The nanoarray structures present on the surfaces of some insect wings such as those of the cicada and dragonfly afford the insect antireflective properties, which can assist in protecting them from predators (Watson et al. 2008). The superhydrophobic and self-cleaning properties can also assist in keeping their surfaces clean and free from contaminants that may also adversely impact their antireflective properties. The self-cleaning properties of these insect wings can be further enhanced due to the presence of turbulent conditions during their flight (Nishimoto and Bhushan 2013).

2.3.4 Superoleophobicity

Superhydrophobicity is the key for terrestrial organisms to deal with contaminants, however it is not a practical option for aquatic organisms, since their living conditions require constant contact with water. In order to cope with this difference in living conditions, nature has employed a different, but similar concept. The surfaces of these organisms are modified so that their surfaces remain wet but unable to be wet by oils, the main source of contaminants, particularly with modern types of marine pollution. Several aquatic species exhibit superoleophobicity rather than superhydrophobicity, exhibiting oil contact angles (OCA) greater than 150° when submerged in water. These organisms possess hierarchical surface structures that are self-cleaning, antifouling and promote low-drag conditions when moving through water (Bixler and Bhushan 2013).

For example, dolphin (Fish and Hui 1991), whale (Baum et al. 2002) and shark skin is known to reduce drag and improve fouling resistance. The skin of bottlenose dolphins *Tursiops truncatus* and the killer whale *Orcinus orca* are covered by dermal ridges positioned such that they are transverse to the direction of flow (Ridgway and Carder 1993; Fish 2006). Another whale, *Globicephala melas* has enclosed nanopores on their patterned ridges, which exhibit great antifouling ability (Baum et al. 2002). Shark skin is covered by dermal denticles shaped like small ribs (or 'riblets'). The denticles are oriented so that they align with the direction of fluid flow as the shark swims through the water. The low drag riblet microstructure, together with a mucous layer on the surface, allows the shark to remain flexible and clean (Bushnell and Moore 1991; Bechert et al. 1997; Dean and Bhushan 2010). This surface structure also provides protection from abrasion, which in turn minimises the opportunities for microorganisms to adhere (Bhushan 2012). Fish scales are another example of self-cleaning surfaces in aquatic environments (Hay 1996). They perform in a manner that is very similar to the shark skin. Their surfaces are covered by sector-like scales (diameter of 4–5 mm), which are covered by papillae (100–300 μm in length and 30–40 μm in width), and exhibit a particularly high oil contact angle in water (163°).

The surface structures of snail shells have been commercially exploited in the construction of snail shell-inspired self-cleaning surfaces for outdoor walls (Nishimoto and Bhushan 2013). These surfaces exhibit the ability to remain clean, despite their dwelling environment and their appearance on rainy days. The surface of snail shells is comprised of a regularly rough structure consisting of line grooves (pitch of 0.5 mm), smaller grooves crossing the line groove (pitch of 0.1 mm) and micro-grooves between the line grooves (pitch of 10 μm). The surface of snail shells is covered by a regular hierarchical structure that ranges in size from micrometers to millimeters, which may facilitate water entrapment. Compared to superhydrophobic surfaces, which entrap air within their hierarchical structure, superoleophobic surfaces trap water molecules. This water-entrapment system helps the shells remain wet, yet remain clean under their semi-aquatic living conditions. This is a key factor that contributes to their ability to self-clean, in the way that their usually wetted surface is rarely able to be contaminated (Nishimoto and Bhushan 2013).

2.4 Mechanobactericidal Activity

The inspiration that can be obtained from insects appears to be almost unlimited. Ivanova et al. recently found that the robust hexagonal arrays of 'nanopillars' on the surfaces of *Psaltoda claripennis* cicada wings are bactericidal (Ivanova et al. 2012). This nanopattern present on the wing surfaces penetrated attaching *Pseudomonas*

Fig. 2.5 Bacterial cells were found to be killed by the physical action of the surface of (**a**, **a1**) cicada wings *Psaltoda claripennis*, and (**b-b3**) dragonfly wings *Diplacodes bipunctata*

aeruginosa cells, killing them with extreme efficiency (Fig. 2.5a, a1). The surface of the cicada wings retained its lethality against these Gram negative pathogenic bacteria even after the surface was coated with a 10 nm-thick layer of gold, which indicated that the bactericidal properties of the cicada wing surfaces arose from the physical properties of the wing surfaces, rather than from their chemical composition. It was also reported that the wings consistently killed other Gram-negative bacteria, i.e., *Branhamella catarrhalis*, *E. coli*, and *Pseudomonas fluorescens*, however Gram-positive cells (*Bacillus subtilis*, *Planococcus maritimus*, and *Staphylococcus aureus*) were found to be resistant to the action of the wing surface (Hasan et al. 2012). Cicada wings were the first example of a surface with bactericidal properties that arose as a result of purely physical action.

To explain this phenomenon, biophysical models were constructed to describe the interaction taking place between the bacterial cells and the nanopatterns present on the surface of the cicada wings (Pogodin et al. 2013). Mathematical calculations revealed that the nanopillars did not pierce the cells but rather the cells were stretched in the regions between the nanopillars as they adsorbed onto the wing surface, until the point of cell rupture. It was also found that the more rigid the cell membrane, the harder they were to break, which was consistent with the results obtained for the Gram-positive bacteria that attached to the wing surface, but were not killed by the action of the nanopillars; the thicker layer of peptidoglycan present in the cell wall afforded the cells a greater dgree of rigidity, making them resistant to the action of the wing nanopillars. This was supported experimentally by decreasing the rigidity of Gram-positive cells though microwave treatment. *B. subtilis*, *S. aureus*, and *Planococcus maritimus* were used as bacterial species. After microwave treatment, all three bacterial species showed a high level of susceptibility to the action of the cicada wing surfaces.

In contrast to cicada wings, which only showed effectiveness against Gram negative bacteria, the surfaces of dragonfly wings were shown to have the ability to kill a large range of bacterial species, including Gram-negative (*Pseudomonas aeruginosa*), and Gram-positive, (*Staphylococcus aureus* and *Bacillus subtilis*) bacteria and even endospores. Similar to cicada wings, dragonfly wings surfaces are covered by a layer of nanopillar-like structures, which punctured all types of bacterial cells that came into contact with the surface, as demonstrated in Fig. 2.5, b1–3. A synthetic material known as black silicon that mimics the surface structure of these dragonfly wings also demonstrated antibacterial properties against these different types of bacterial cells (Ivanova et al. 2013a). The discovery of the bactericidal properties possessed by these insect wings has brought them into focus as promising new prospects as templates for the production of synthetic biocidal surfaces.

References

Angst EC (1923) The fouling of ships bottoms by bacteria. Rep., Bur. Constr. Repair. US Navy Department, Washington, DC

Arsene C, Schulz S, Van Loon JJA (2002) Chemical polymorphism of the cuticular lipids of the cabbage white *Pieris rapae*. J Chem Ecol 28(12):2627–2631

Barthlott W, Neinhuis C (1997) Purity of the sacred lotus, or escape from contamination in biological surfaces. Planta 202(1):1–8

Barthlott W, Neinhuis C, Cutler D, Ditsch F, Meusel I, Theisen I, Wilhelmi H (1998) Classification and terminology of plant epicuticular waxes. Bot J Linn Soc 126(3):237–260

Baum C, Meyer W, Stelzer R, Fleischer LG, Siebers D (2002) Average nanorough skin surface of the pilot whale (*Globicephala melas*, Delphinidae): considerations on the self-cleaning abilities based on nanoroughness. Mar Biol 140(3):653–657

Bechert DW, Bruse M, Hage W, Van Der Hoeven JGT, Hoppe G (1997) Experiments on drag-reducing surfaces and their optimization with an adjustable geometry. J Fluid Mech 338:59–87

Behlau I, Gilmore MS (2008) Microbial biofilms in ophthalmology and infectious disease. Arch Ophthalmol 126(11):1572–1581

Bhushan B (2012) Bioinspired structured surfaces. Langmuir 28(3):1698–1714

Bhushan B, Jung YC, Koch K (2009) Self-cleaning efficiency of artificial superhydrophobic surfaces. Langmuir 25(5):3240–3248

Bixler GD, Bhushan B (2013) Fluid drag reduction and efficient self-cleaning with rice leaf and butterfly wing bioinspired surfaces. Nanoscale 5(17):7685–7710

Boeve JL, Voigt D, Gorb SN (2011) Crystalline wax coverage of the cuticle in easy bleeding sawfly larvae. Arthropod Struct Dev 40(2):186–189

Bos R, Van Der Mei HC, Busscher HJ (1999) Physico-chemistry of initial microbial adhesive interactions – its mechanisms and methods for study. FEMS Microbiol Rev 23(2):179–229

Buckner JS (2010) Oxygenated derivatives of hydrocarbons. In: Blomquist GJ, Bagnères A-G (eds) Insect hydrocarbons: biology, biochemistry, and chemical ecology. Cambridge University Press, Cambridge, pp 187–203

Bulgarelli D, Schlaeppi K, Spaepen S, Van Themaat EVL, Schulze-Lefert P (2013) Structure and functions of the bacterial microbiota of plants. Annu Rev Plant Biol 64:807–838

Buschhaus C, Jetter R (2011) Composition differences between epicuticular and intracuticular wax substructures: how do plants seal their epidermal surfaces? J Exp Bot 62(3):841–853
Buschhaus C, Herz H, Jetter R (2007) Chemical composition of the epicuticular and intracuticular wax layers on adaxial sides of *Rosa canina* leaves. Ann Bot 100(7):1557–1564
Bushnell DM, Moore KJ (1991) Drag reduction in nature. Annu Rev Fluid Mech 23(1):65–79
Byun D, Hong J, Saputra KJH, Lee YJ, Park HC, Byun BK, Lukes JR (2009) Wetting characteristics of insect wing surfaces. J Bionic Eng 6(1):63–70
Cassie ABD, Baxter S (1944) Wettability of porous surfaces. Trans Faraday Soc 40:546–551
Dalton HM, March PE (1998) Molecular genetics of bacterial attachment and biofouling. Curr Opin Biotechnol 9(3):252–255
Daniels R, Vanderleyden J, Michiels J (2004) Quorum sensing and swarming migration in bacteria. FEMS Microbiol Rev 28(3):261–289
Dean B, Bhushan B (2010) Shark-skin surfaces for fluid-drag reduction in turbulent flow: a review. Philos Trans A Math Phys Eng Sci 368(1929):4775–4806
Di Iaconi C, Ramadori R, Lopez A, Passino R (2005) Hydraulic shear stress calculation in a sequencing batch biofilm reactor with granular biomass. Environ Sci Technol 39(3):889–894
Ensikat HJ, Ditsche-Kuru P, Neinhuis C, Barthlott W (2011) Superhydrophobicity in perfection: the outstanding properties of the lotus leaf. Beilstein J Nanotechnol 2(1):152–161
Fish FE (2006) The myth and reality of Gray's paradox: implication of dolphin drag reduction for technology. Bioinspir Biomim 1(2):R17–R25
Fish FE, Hui CA (1991) Dolphin swimming – a review. Mammal Rev 21(4):181–195
Franklin NM, Rogers NJ, Apte SC, Batley GE, Gadd GE, Casey PS (2007) Comparative toxicity of nanoparticulate ZnO, bulk ZnO, and $ZnCl_2$ to a freshwater microalga (*Pseudokirchneriella subcapitata*): the importance of particle solubility. Environ Sci Technol 41(24):8484–8490
Fu G, Vary PS, Lin CT (2005) Anatase TiO_2 nanocomposites for antimicrobial coatings. J Phys Chem B 109(18):8889–8898
Garrett TR, Bhakoo M, Zhang Z (2008) Bacterial adhesion and biofilms on surfaces. Prog Nat Sci 18(9):1049–1056
Gazit E (2007) Self-assembled peptide nanostructures: the design of molecular building blocks and their technological utilization. Chem Soc Rev 36(8):1263–1269
Geesey GG, Richardson WT, Yeomans HG, Irvin RT, Costerton JW (1977) Microscopic examination of natural sessile bacterial populations from an alpine stream. Can J Microbiol 23(12):1733–1736
Gelover S, Gómez LA, Reyes K, Leal MT (2006) A practical demonstration of water disinfection using TiO_2 films and sunlight. Water Res 40(17):3274–3280
Gorb SN (1999) Serial elastic elements in the damselfly wing: mobile vein joints contain resilin. Naturwissenschaften 86(11):552–555
Grimaldi D, Engel MS (2005) Insects take to the skies. In: Evolution of the insects. Cambridge University Press, New York, pp 155–178
Guo Z, Liu W (2007) Biomimic from the superhydrophobic plant leaves in nature: binary structure and unitary structure. Plant Sci 172(6):1103–1112
Hasan J, Webb HK, Truong VK, Pogodin S, Baulin VA, Watson GS, Watson JA, Crawford RJ, Ivanova EP (2012) Selective bactericidal activity of nanopatterned superhydrophobic cicada *Psaltoda claripennis* wing surfaces. Appl Microbiol Biotechnol 97(20):1–6
Hay ME (1996) Marine chemical ecology: what's known and what's next? J Exp Mar Biol Ecol 200(1–2):103–134
Henrici AT (1933) Studies of freshwater bacteria. I. A direct microscopic technique. J Bacteriol 25(3):277–287
Holmes A, Doré CJ, Saraswatula A, Bamford KB, Richards MS, Coello R, Modi N (2008) Risk factors and recommendations for rate stratification for surveillance of neonatal healthcare-associated bloodstream infection. J Hosp Infect 68(1):66–72
Hsiao MT, Chen SF, Shieh DB, Yeh CS (2006) One-pot synthesis of hollow Au_3Cu_1 spherical-like and biomineral botallackite $Cu_2(OH)_3Cl$ flowerlike architectures exhibiting antimicrobial activity. J Phys Chem B 110(1):205–210

Ibáñez JA, Litter MI, Pizarro RA (2003) Photocatalytic bactericidal effect of TiO_2 on *Enterobacter cloacae*. Comparative study with other Gram (-) bacteria. J Photochem Photobiol A Chem 157(1):81–85

In S, Orlov A, Berg R, García F, Pedrosa-Jimenez S, Tikhov MS, Wright DS, Lambert RM (2007) Effective visible light-activated B-doped and B, N-codoped TiO2 photocatalysts. J Am Chem Soc 129(45):13790–13791

Ivanova EP, Hasan J, Webb HK, Truong VK, Watson GS, Watson JA, Baulin VA, Pogodin S, Wang JY, Tobin MJ, Löbbe C, Crawford RJ (2012) Natural bactericidal surfaces: mechanical rupture of *Pseudomonas aeruginosa* cells by cicada wings. Small 8(16):2489–2494

Ivanova EP, Hasan J, Webb KW, Gervinskas G, Juodkazis S, Truong VK, Wu AHF, Lamb RN, Baulin VA, Watson GS, Watson JA, Mainwaring DE, Crawford RJ (2013a) Bactericidal activity of black silicon. Nat Commun 4:2838–2844

Ivanova EP, Nguyen SH, Webb HK, Hasan J, Truong VK, Lamb RN, Duan X, Tobin MJ, Mahon PJ, Crawford RJ (2013b) Molecular organization of the nanoscale surface structures of the dragonfly *Hemianax papuensis* wing epicuticle. PLoS ONE 8(7):e67893

Jeffree CE (2006) In: Riederer M, Müller C (eds) Biology of the plant cuticle. Blackwell Publishing, Oxford, pp 11–125

Jetter R, Kunst L (2008) Plant surface lipid biosynthetic pathways and their utility for metabolic engineering of waxes and hydrocarbon biofuels. Plant J 54(4):670–683

Jetter R, Schäffer S, Riederer M (2000) Leaf cuticular waxes are arranged in chemically and mechanically distinct layers: evidence from *Prunus laurocerasus L.* Plant Cell Environ 23(6):619–628

Jones N, Ray B, Ranjit KT, Manna AC (2008) Antibacterial activity of ZnO nanoparticle suspensions on a broad spectrum of microorganisms. FEMS Microbiol Lett 279(1):71–76

Jung YC, Bhushan B (2006) Contact angle, adhesion and friction properties of micro- and nanopatterned polymers for superhydrophobicity. Nanotechnology 17(19):4970–4980

Kenrick P, Crane PR (1997) The origin and early evolution of plants on land. Nature 389(6646):33–39

Kjelleberg S, Molin S (2002) Is there a role for quorum sensing signals in bacterial biofilms? Curr Opin Microbiol 5(3):254–258

Klevens RM, Edwards JR, Richards CL Jr, Horan TC, Gaynes RP, Pollock DA, Cardo DM (2007) Estimating health care-associated infections and deaths in U.S. Hospitals, 2002. Public Health Rep 122(2):160–166

Koch K, Ensikat HJ (2008) The hydrophobic coatings of plant surfaces: epicuticular wax crystals and their morphologies, crystallinity and molecular self-assembly. Micron 39(7):759–772

Koch K, Dommisse A, Barthlott W (2006) Chemistry and crystal growth of plant wax tubules of lotus (*Nelumbo nucifera*) and nasturtium (*Tropaeolum majus*) leaves on technical substrates. Cryst Growth Des 6(11):2571–2578

Koch K, Bhushan B, Ensikat HJ, Barthlott W (2009) Self-healing of voids in the wax coating on plant surfaces. Philos Trans A Math Phys Eng Sci 367(1894):1673–1688

Kreuz P, Arnold W, Kesel AB (2001) Acoustic microscopic analysis of the biological structure of insect wing membranes with emphasis on their waxy surface. Ann Biomed Eng 29(12):1054–1058

Lichter JA, Van Vlietpa KJ, Rubner MF (2009) Design of antibacterial surfaces and interfaces: polyelectrolyte multilayers as a multifunctional platform. Macromolecules 42(22):8573–8586

Liu Y, Tay JH (2002) The essential role of hydrodynamic shear force in the formation of biofilm and granular sludge. Water Res 36(7):1653–1665

Lockey KH (1980) Insect cuticular hydrocarbons. Comp Biochem Physiol B: Biochem Mol Biol 65(3):457–462

Lockey KH (1985) Insect cuticular lipids. Comp Biochem Physiol B: Biochem Mol Biol 81(2):263–273

Lockey KH (1988) Lipids of the insect cuticle: origin, composition and function. Comp Biochem Physiol B: Biochem Mol Biol 89(4):595–645

Maness PC, Smolinski S, Blake DM, Huang Z, Wolfrum EJ, Jacoby WA (1999) Bactericidal activity of photocatalytic TiO_2 reaction: toward an understanding of its killing mechanism. Appl Environ Microbiol 65(9):4094–4098

Moussian B (2010) Recent advances in understanding mechanisms of insect cuticle differentiation. Insect Biochem Mol Biol 40(5):363–375

Neinhuis C, Barthlott W (1997) Characterization and distribution of water-repellent, self-cleaning plant surfaces. Ann Bot 79(6):667–677

Nelson DR, Blomquist LG (1995) Insect waxes. In: Hamilton RJ, Christie WW (eds) Waxes: chemistry, molecular biology and functions. The Oily Press Ltd Hamilton RJ, Dundee, pp 1–90

Nelson DR, Charlet LD (2003) Cuticular hydrocarbons of the sunflower beetle, *Zygogramma exclamationis*. Comp Biochem Physiol B: Biochem Mol Biol 135(2):273–284

Nguyen SH, Webb HK, Hasan J, Tobin MJ, Crawford RJ, Ivanova EP (2013) Dual role of outer epicuticular lipids in determining the wettability of dragonfly wings. Colloids Surf B: Biointerfaces 106:126–134

Nishimoto S, Bhushan B (2013) Bioinspired self-cleaning surfaces with superhydrophobicity, superoleophobicity, and superhydrophilicity. RSC Adv 3(3):671–690

Nosonovsky M, Bhushan B (2005) Roughness optimization for biomimetic superhydrophobic surfaces. Microsyst Technol 11(7):535–549

Nosonovsky M, Bhushan B (2007) Multiscale friction mechanisms and hierarchical surfaces in nano- and bio-tribology. Mater Sci Eng R Rep 58(3–5):162–193

Nosonovsky M, Bhushan B (2008) Multiscale dissipative mechanisms and hierarchical surfaces: friction, superhydrophobicity and biomimetics. Springer, Berlin

O'Toole G, Kaplan HB, Kolter R (2000) Biofilm formation as microbial development. Annu Rev Microbiol 54:49–79

Pogodin S, Hasan J, Baulin VA, Webb HK, Truong VK, Nguyen THP, Boshkovikj V, Fluke CJ, Watson GS, Watson JA, Crawford RJ, Ivanova EP (2013) Biophysical model of bacterial cell interactions with nanopatterned cicada wing surfaces. Biophys J 104(4):835–840

Qi L, Xu Z, Jiang X, Hu C, Zou X (2004) Preparation and antibacterial activity of chitosan nanoparticles. Carbohydr Res 339(16):2693–2700

Rai M, Yadav A, Gade A (2009) Silver nanoparticles as a new generation of antimicrobials. Biotechnol Adv 27(1):76–83

Ramsay JA (1935) The evaporation of water from the cockroach. J Exp Biol 12:373–383

Ridgway SH, Carder DA (1993) Features of dolphin skin with potential hydrodynamic importance. IEEE Eng Med Biol Manag 12(3):83–88

Ryu JH, Beuchat LR (2005) Biofilm formation and sporulation by *Bacillus cereus* on a stainless steel surface and subsequent resistance of vegetative cells and spores to chlorine, chlorine dioxide, and a peroxyacetic acid-based sanitizer. J Food Protect 68(12):2614–2622

Samuels L, Kunst L, Jetter R (2008) Sealing plant surfaces: cuticular wax formation by epidermal cells. Annu Rev Plant Biol 59:683–707

Schmidt S, Winter J, Gallert C (2012) Long-term effects of antibiotics on the elimination of chemical oxygen demand, nitrification, and viable bacteria in laboratory-scale wastewater treatment plants. Arch Environ Contam Toxicol 63(3):354–364

Schultz MP, Bendick JA, Holm ER, Hertel WM (2011) Economic impact of biofouling on a naval surface ship. Biofouling 27(1):87–98

Seo YS, Lee DY, Rayamahji N, Kang ML, Yoo HS (2008) Biofilm-forming associated genotypic and phenotypic characteristics of *Staphylococcus* spp. isolated from animals and air. Res Vet Sci 85(3):433–438

Simões M, Simões LC, Vieira MJ (2009) Species association increases biofilm resistance to chemical and mechanical treatments. Water Res 43(1):229–237

Su Y, Ji B, Zhang K, Gao H, Huang Y, Hwang K (2010) Nano to micro structural hierarchy is crucial for stable superhydrophobic and water-repellent surfaces. Langmuir 26(7):4984–4989

Van Maarseveen C, Han H, Jetter R (2009) Development of the cuticular wax during growth of *Kalanchoe daigremontiana* (Hamet et Perr. de la Bathie) leaves. Plant Cell Environ 32(1):73–81

Wan Y, Cong Q, Wang X, Yan Z (2008) The wettability and mechanism of geometric non-smooth structure of dragonfly wing surface. J Bionic Eng 5(SUPPL):40–45

Watson GS, Myhra S, Cribb BW, Watson JA (2008) Putative functions and functional efficiency of ordered cuticular nanoarrays on insect wings. Biophys J 94(8):3352–3360

Webb HK, Hasan J, Truong VK, Crawford RJ, Ivanova EP (2011) Nature inspired structured surfaces for biomedical applications. Curr Med Chem 18(22):3367–3375

Wenzel RN (1949) Surface roughness and contact angle. J Phys Colloid Chem 53(9):1466–1467

Wootton RJ (1992) Functional morphology of insect wings. Annu Rev Entomol 37(1):113–140

Wu MY, Sendamangalam V, Xue Z, Seo Y (2012) The influence of biofilm structure and total interaction energy on *Escherichia coli* retention by *Pseudomonas aeruginosa* biofilm. Biofouling 28(10):1119–1128

Yan YY, Gao N, Barthlott W (2011) Mimicking natural superhydrophobic surfaces and grasping the wetting process: a review on recent progress in preparing superhydrophobic surfaces. Adv Colloid Interface Sci 169(2):80–105

Yasuyuki M, Kunihiro K, Kurissery S, Kanavillil N, Sato Y, Kikuchi Y (2010) Antibacterial properties of nine pure metals: a laboratory study using *Staphylococcus aureus* and *Escherichia coli*. Biofouling 26(7):851–858

Young T (1805) An essay on the cohesion of fluids. Philos Trans R Soc Lond 95:65–87

Zhang X, Wang L, Levänen E (2013) Superhydrophobic surfaces for the reduction of bacterial adhesion. RSC Adv 3(30):12003–12020

Chapter 3
Artificial Antibacterial Surfaces that are Simple to Fabricate

Armandas Balčytis, Gediminas Seniutinas, Florian Lapierre, and Saulius Juodkazis

Abstract Nanotexturing to produce surfaces containing hydrophobic functionality was demonstrated with the production of plasma etched silicon (black silicon, bSi), chemically etched copper (black copper, bCu), and nano-wrinkled metal films formed on thermally relaxed, or shrunken plastic sheets. The features common to all of these surfaces high surface area, a surface structure resulting from self-organization and are surfaces that can be coated by metallic films using sputtering techniques. Such surfaces are applicable for sensor applications. The potential for applying these surfaces in anti-biofouling and hydrophobic applications is discussed.

Keywords Biomimetic surfaces • Nanotexturing • Wettability • Hydrophobicity • Fabrication • Black silicon

A. Balčytis (✉)
Centre for Micro-Photonics, Faculty of Science, Engineering and Technology, Swinburne University of Technology, Hawthorn, VIC 3122, Australia

Melbourne Centre for Nanofabrication (MCN), Australian National Fabrication Facility (ANFF), Clayton, VIC 3168, Australia

Center for Physical Sciences and Technology, A. Goštauto 9, LT-01108 Vilnius, Lithuania

G. Seniutinas • S. Juodkazis
Centre for Micro-Photonics, Faculty of Science, Engineering and Technology, Swinburne University of Technology, Hawthorn, VIC 3122, Australia

Melbourne Centre for Nanofabrication (MCN), Australian National Fabrication Facility (ANFF), Clayton, VIC 3168, Australia

F. Lapierre
Micro Nano Research Laboratory, RMIT, VIC 3101 Melbourne, Australia

E.P. Ivanova, R.J. Crawford (eds.), *Antibacterial Surfaces*,
DOI 10.1007/978-3-319-18594-1_3

3.1 Introduction

The art of artificially imitating the mechanisms that occur in nature in order to integrate them into technological domains, or 'Biomimetics' has attracted a large amount of interest in the past decade. One can refer to, for example, the self-cleaning property of lotus leaves (*Nelumbo nucifera*), which allows them to stay clean, or to the antibacterial properties of the wings of butterflies, dragonflies or cicadas (*Psaltoda claripennis*) as natural systems with inspirational surfaces (Hasan et al. 2012; Tobin et al. 2013). The unique behaviour of such natural surfaces has been linked to the mechanisms that take place between the interacting medium and the micro-nanotextured surface of the substrate. The comprehension of these mechanisms, together with the ability to replicate such surfaces using various micro-nanotechnology platforms, has meant that these surfaces could be exploited in the design of, for example, microfluidic devices and in other biotechnological applications. The justification for integrating such surfaces into microfluidic devices is based on several advantages associated with such surfaces; there is a reduction of energy needed for the motion of objects over these surfaces due to their low friction, negligible amounts of surface pollution being present, and the presence of a high gas exchange ratio with liquids in contact with the surface.

The fabrication of substrates with large surface areas containing micro-nanotexturation has always been the focus of basic research. Parallel surface processing methods such as chemical and plasma etching as well as dewetting or thermal treatment represent preferable techniques for the mass production of such surfaces. Nowadays, however, serial surface modification methods such as laser ablation and the formation of ripples are gaining a great deal of interest as they present competitive fabrication rates due to advances in fast substrate stages and laser beam scanning technologies.

Textured surfaces can be prepared as a mould for further replication via thermal molding directly onto a substrate surface, or via a nickel shim copying process. Roll-to-roll processing has the potential to be a very efficient method for the mass production of nanotextured surfaces, however, there is difficulty experienced in replicating surfaces that contain surface features in the order of tens-of-nanometers, particularly when the aspect ratio is high, meaning the feature *height/width* ratio is greater than 2. Hence, the preparation of surfaces by simple and fast processing techniques will be in high demand. The specific anti-bacterial properties of nanotextured surfaces is the focus of this chapter. Surfaces can be called anti-bacterial under certain circumstances; (i) when the attachment of bacteria is hampered due to its physical or chemical morphology or composition, (ii) the surface is antibacterial because of its hydrophobicity or hydrophilicity, or (iii) when bacteria are killed when coming into contact with the nano-textured surface. All three properties can prevent the formation of a biofilm. Such substrates are of great benefit when fabricating anti-bio-fouling surfaces in food, medical, and industrial fields. Another aspect of such surfaces can be demonstrated with their use as sensors, especially for surface enhanced Raman scattering (SERS) spectroscopy applications and in the detection of molecules at very low concentrations.

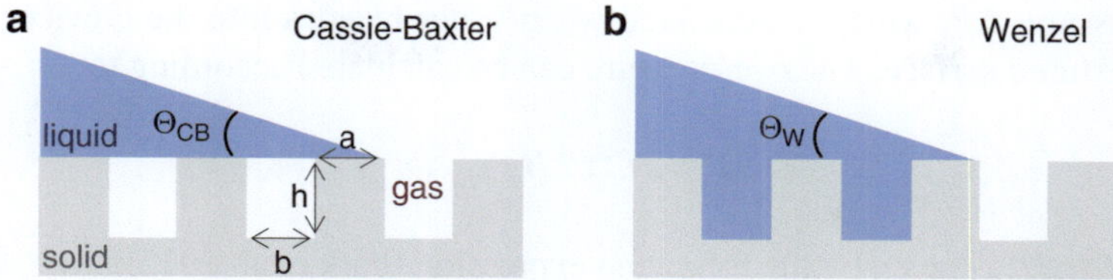

Fig. 3.1 Cassie-Baxter and Wenzel wetting conditions. Geometric features on the nano-micro-scale *a, b, h*, their shape, and ordering are underpinning the wettability of the surface as defined by the contact angle θ

Here, we focus on several particular cases of surface nanotexturing that can be reproduced over large areas, in a comparatively simple way, using parallel rather than serial processing techniques, e.g., laser ablation. Such surfaces can adopt both the Cassie-Baxter and Wenzel wetting regimes (Fig. 3.1). In addition to the surface topology derived bactericidal properties, nano-textured materials (Ivanova et al. 2013) have applications in SERS sensing. The ability to simply fabricate SERS substrates is advantageous, since the substrates currently being used require the use of complex serial high resolution electron or ion beam technologies (Seniutinas et al. 2013; Nishijima et al. 2013, 2014a, b) or laser writing (Chou et al. 2012; Gervinskas et al. 2013a; Jayawardhana et al. 2012; Buividas et al. 2012).

3.2 Wetting

3.2.1 Cassie-Baxter Versus Wenzel State

Nanotextured surfaces usually serve multiple purposes, as they do in nature. For example, the nano-sharp parabolic silica (SiO_2) nanopillars present on modified silica surfaces can serve as an antireflective coating that has a wide angle window function that can be used for the construction of mobile phones or monitors, and due to their hydrophobicity, can be used for solar cell applications (Park et al. 2012). The hydrophobicity of nano-textured surfaces can occur naturally due to the surface energy of these materials, but it can be further enhanced by chemically treating the surface through, for example, the deposition of self-assembled molecular films. Such surfaces trap air bubbles within the surface, making it even more hydrophobic through the surface entering the Cassie- Baxter wetting regime. Even in cases where the nano-textured surface is hydrophilic, as is the case for the SiO_2 covering black silicon (due to the native 2 nm thick SiO_2) the air pockets that are trapped below any water droplets coming in contact with the surface make it hydrophobic.

The robustness of the nano-textured surface against the transformation from the Cassie-Baxter state into the Wenzel wetting regime depends on the magnitude of the overpressure, ΔP, which is generated when water recedes into the crevices in the nano-textured surface. The overpressure can be calculated according to:

$$\Delta P = 2T(z/H)\gamma_{LV}/l_{cap} \tag{3.1}$$

where $T(z/H)$ is given by the following expression (Park et al. 2012):

$$T(z/H) = -\frac{\pi l_{cap} \frac{W(z/H)}{P}\cos(\theta-\phi)}{P\left(1-\pi\left(\frac{W(z/H)}{P}\right)^2\right)} \tag{3.2}$$

where γ_{LV} is the water surface tension, θ is the advancing contact angle of the water front on the surface (see inset in Fig. 3.2), ϕ is the tip angle of a parabolic nanopillar (for such nanopillars, $W(z/H) = P\sqrt[3]{(z/H)/2}$), P is the period (distance) between the neighboring nanopillars, $l_{cap} = \sqrt{\gamma LV / g\rho}$ is the capillary length with ρ being the density of water and g being the acceleration due to gravity. It should be noted that Eq. 3.2 is general, and that the actual shape of the nanopillars is taken into account by the $W(z/H)$ function. ΔP is positive regardless of whether the surface is hydrophilic or hydrophobic, and hence the sign of $\cos(\theta)$ should be carefully considered when using Eq. 3.2.

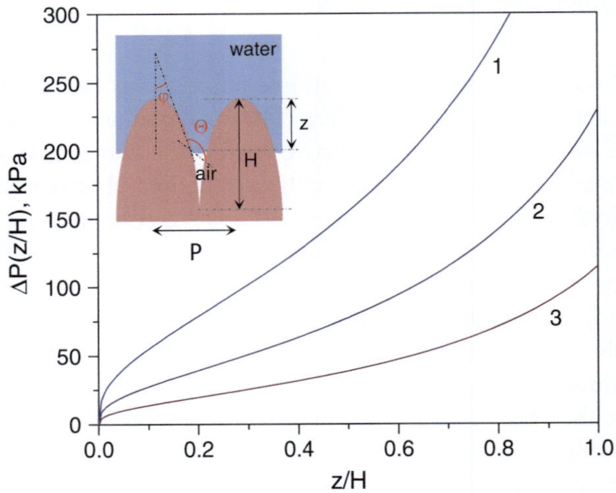

Fig. 3.2 Pressure of trapped air when water is placed on a hydrophobic surface resulting in a contact angle of $\theta = 100°$ on parabolic nanopillars, as calculated by Eq. 3.1 (Park et al. 2012). Other parameters used for these calculations are: $\phi = 5°$, the surface tension of water (γ_{LV}) is 72 mN/m, the period between nanopillars (P) is 200 nm (1), 400 nm (2), and 800 nm (3)

Figure 3.2 shows that the pressure builds up as water recedes into the structure from, for example, the pressure of the incoming droplet, which is a function of the different periods of the nanopillar-like structure. The smallest period $P = 200$ nm, used for preparing anti-reflective and wide- angle screens (Park et al. 2012), has been found to be the most resistant to the surface entering into the Wenzel state (destroying air pockets required for the Cassie-Baxter regime), which occurs when water reaches the bottom of the structure $z/H = 1$. The calculations for 400 nm nanopillar separation are applicable to black silicon samples, on which a contact angle of 100° was observed without any hydrophobicity inducing surface treatment (the same wetting angle was used for calculations shown in Fig. 3.2).

By changing the advancing angle to that corresponding to a partial wetting condition ($\theta < 90°$) there is no significant change in the ΔP behaviour (not shown). In such cases, air is trapped below the surface and therefore, qualitatively speaking, the same behaviour would occur under receding water conditions. The smaller the period, the higher the ΔP pressures that would be generated, which favours the condition of surface hydrophobicity via the surface entering the Cassie-Baxter regime. Pyramidal nanopillars that are typically found on the surface of black silicon behave in a similar fashion to that of the simulated parabolic nanopillars, with a *height/width* aspect ratio of 5.5 (Park et al. 2012).

The repellency of water is an important feature of nano-textured surfaces in naturally occurring systems. The water hammer pressure from a droplet falling onto a nano-textured surface, P_{WH} *is approximately* 0.5 ρcv, where c is the speed of sound in water and v is the velocity of the droplet. It is the highest dynamic mechanical pressure that the nanopillars experience under natural environmental exposure conditions, for example, in the case of cicada wings. The pressure P_{WH} decays to the Bernoulli pressure $P_B \sim \rho v^2$. Triangular nanopillars are more mechanically robust against fracture as well as for the generation of larger ΔP values (Fig. 3.2), preventing the surface from entering the Wenzel state, which would allow water to enter into the array of nanopillars on the surface of the wings of dragon flies and cicadas, considerably increasing the weight being borne by the wing. During surface drying, the capillary forces can be high for the wetting liquid, and sufficiently large to break the nano-structures, as was observed for laser polymerised 3D periodic patterns with sub-micrometer feature sizes (Kondo et al. 2005). In addition to the structural features, the wings of cicadas are coated with epicuticular waxes that makes them more hydrophobic, and therefore more resistant to water entering the surface structure (Tobin et al. 2013).

3.2.2 Surface Treatment to Impart Hydrophobicity

In order to impart hydrophobic properties to the surfaces of black silicon and black copper, these substrates were subjected to treatment with trichloro(1H,1H,2H,2H-perfluorooctyl)silane (PFTS) under desiccator or nitrogen glovebox conditions, with evaporation taking place for several hours in a low (10^{-3}) bar vacuum conditions. This process deposits randomly oriented and self-assembled molecular monolayers (SAM) of PFTS onto the surface (Lapierre et al. 2011a).

3.3 Fabrication of Nanotextured Surfaces

3.3.1 Black Silicon

Large area plasma etching techniques can be used for the nanotexturing of surfaces. Despite requiring a vacuum, this technique is simple, efficient and practical, because by combining reactive ion and inductively coupled plasma etching (RIE and ICP), rapid surface processing can take place at etch rates of tens-of-nm per minute over wafer-sized substrate areas. The etched patterns can be dened lithographically or by using self-organized etching masks. A known example of the latter case is in the formation of black silicon, where the initial steps of native oxide removal in the plasma creates a random surface mask that eventually results in the formation of the characteristic sharp nanopillars found on black silicon. Using RIE and ICP, aspect ratios exceeding 10 can be reached for the nano-features on the surface of dielectric materials and semiconductors.

The plasma etching and fabrication of black silicon was the first step in the study reported here. Subsequent methods employed electrochemical anodisation to change the morphology and pattern of the black silicon. In this case, back side illumination was employed to define the region for anodization.

As fabricated, black silicon is hydrophobic, exhibiting a water contact angle of 100°, provided the surface is covered by very high aspect ratio nanopillars. The highest aspect ratio that was fabricated was approximately 10. Modification of the black silicon surface to modify its hydrophobicity can be done by evaporation of PFTS onto the surface under controlled conditions.

Photo-electrochemical etching of the surface was controlled using back-side illumination. In this case, a projection of the illuminated area was etched at the interface between the silicon and an aqueous HF solution. Etching relies on the supply of holes (h) to the liquid–solid interface while the presence of electrons (e) inhibits the etching process. Any holes present will diffuse and drift along the potential lines between the electrodes, away from the absorption region on the back-side face where ($e-h$) pairs are generated (see, Fig. 3.3). n-type silicon is required for the efficient delivery of photo-generated holes to the etching surface (Juodkazis et al. 2013). In this case, 0.4-mm-thick Si wafers were used to prepare black silicon that was covered by nanopillars with aspect ratios of approximately 2, using procedures reported elsewhere (Žukauskas et al. 2013). When black silicon is immersed into a HF solution, both the dissolution of silica and oxidation of the silicon takes place, however, it is possible to control the rate at which these processes take place using an applied bias voltage. A 0.5 V bias relative to the standard hydrogen electrode (RHE) scale was used. A slight variation in the aspect ratio occurs in different black silicon wafers, however, several minutes of etching under exposure to light gives rise to a surface topology that significantly differs from that obtained under the same conditions in the absence of light (Fig. 3.4). Measurement of the anodic current in the absence of light resulted in 0.2 μA erratic jumps, which is related to the oxidation and corrosion of the nanopillars, producing a smoother black silicon sur-

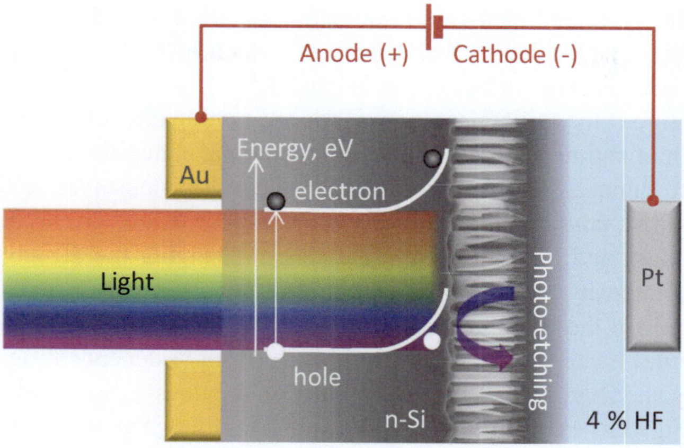

Fig. 3.3 Photo-electrochemical etching of black silicon in 4 % HF solution. The black silicon nanopillars have an aspect ratio of ∼10, up to 2 μm long with a tip apex of 20 nm in diameter; the actual SEM image is incorporated into the schematics of experiments on back-side illumination for controlled wet etching

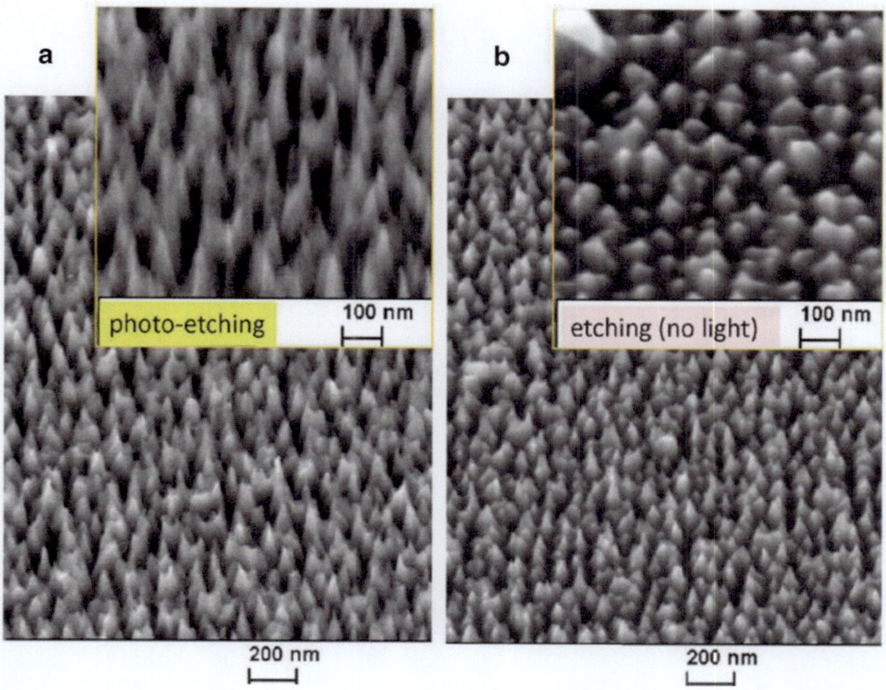

Fig. 3.4 Photo-electrochemical etching of black silicon (Fig. 3.3): after etching with back-side illumination and without it shown in *insets* of (**a**) and (**b**), respectively. The main panels in (**a**) and (**b**) are black silicon surfaces before electrochemical etching has taken place

face. Conversely, under back-side illumination conditions, etching can be further enhanced (via a larger current being present) which increases the aspect ratio of nanopillars covering the surface (Figs. 3.4 and 3.5).

For the p-type black silicon, no photo-induced etching effect was observed under back-side illumination conditions. After HF treatment, nanopillars covering the surface became thinner, however, negligible changes were observed with regard to their aspect ratio, possibly due to the etching rates (corrosion current) being similar under both dark and light exposure conditions. Changing the HF concentration, illumination intensity and spectrum and bias, it is theoretically possible to control the aspect ratio of the nanopillars. This is especially important, since the black silicon surface has bactericidal properties that are strongly dependent on the surface morphology and sharpness of the nanopillars. As previously discussed, larger pressures ΔP (Fig. 3.2) exist for denser arrays of nanopillars; the surface modification technique described above allows the aspect ratio of the nanopillar pattern to be modified without changing the distance between the nanopillars. The back-side illumination-assisted photo-induced etching technique can be used for the directed modification of black silicon patterns.

Fig. 3.5 SEM images of CuO nano-flakes (black copper) sputter coated with Al at different thicknesses (the thickness is indicated for the flat substrate). Black copper was prepared using chemical etching

3.3.2 Black Copper

Plasma etching is significantly less practical for the nano-texturing of metals, which usually make a good mask. Direct chemical or electrochemical etching of metals can be used for the variable scale chemical or/and anodic oxidation of metals. Various nano-textures can be generated using this process. In the case of electrochemical etching, the morphology of the surface pattern, usually with respect to the diameter of the pores, can be adjusted and surfaces with very high aspect ratios can be produced, which are limited by the obtainable rates for delivering the etchant to the surface, and the diffusional removal of the reaction products.

Initially, the as-etched CuO surface does not possess any hydrophobic properties, exhibiting a water contact angle of 3° with extremely high contact angle hysteresis. In order to impart hydrophobic properties to the surface, the CuO substrates were exposed to PFTS (Lapierre et al. 2011a), which substantially increased the contact angle to $160 \pm 2°$ (Fig. 3.6). The surfaces, however, exhibited a reduced contact angle hysteresis of $5 \pm 1°$, which indicated that the liquid/surface interface was not entirely in the Cassie-Baxter state, but instead approaching the Wenzel regime (Lapierre et al. 2010). Water coming into contact with the surface is most likely pinned into the nanotexturation of the CuO flakes; nevertheless, these results showed that CuO surfaces can be made to possess superhydrophobic behaviour, with self-cleaning and hydrophobic properties, both of which are important in solar cell applications. Moreover, high concentrations of an analyte at localized spots, which

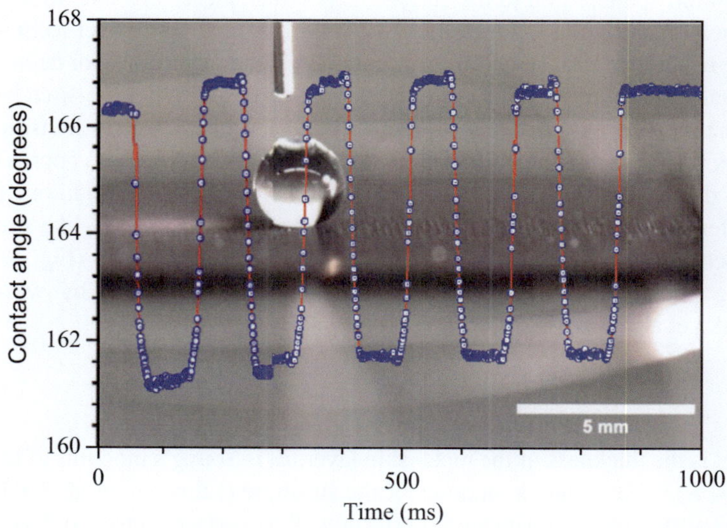

Fig. 3.6 Contact angle hysteresis measurement conducted by drop release onto a 1H,1H,2H,2H-perfluorodecyltrichlorosilane treated CuO-on-Cu substrate. The background shows a photograph depicting the hydrophobic nature of the CuO-on-Cu after surface treatment. Photo *scale bar* is 5 mm

form as the solvent drop dries and recedes on a hydrophobic surface, can strongly enhance SERS signal from a diluted solution. The use of superhydrophobic surfaces in microfluidic systems shows great potential, combining fast and precise biomolecule transportation and manipulation (Lapierre et al. 2011b, 2013), allowing high sensitivity detection (Lapierre et al. 2011c). Integrating such CuO surfaces into bio-micro-electromechanical systems for SERS detection could show great potential.

Higher aspect ratio structures can be created using electrochemical etching processes (Juodkazytė et al. 2013, 2014) with an increase of surface area being monitored by voltammetry (Buividas et al. 2013). CuO flakes with a surface area up to 280 times larger than that of a flat Cu surface were prepared using wet chemical etching techniques (Juodkazytė et al. 2013, 2014). The application of black copper in sensing applications is possible (Song et al. 2010), with the edge of nano-flakes having apex diameters of 10–20 nm, which makes them comparable to those found on the surface of black silicon. The nano-granular pattern resulting from gold and aluminium sputtering on black copper surfaces means that such surfaces would be suitable for SERS applications.

3.3.3 Thermal Shrinkage of Polymer Sheets

The thermal shrinkage of pre-stretched plastic, which arises though a volume phase transition, can be utilised to change the morphology of the patterns formed on both the surface and in the bulk of the polymer (Juodkazis et al. 2004). A nano-thin layer of chromium sputtered on to the surface of polymer sheets was able to be used to form buckled and wrinkled metallic structures. A simple and novel technique for creating nano-structured metallic surfaces involves the physical vapor deposition of thin metal layers onto sheets of commodity polystyrene, or other pre-stretched plastics. When the coated plastic is subsequently heated over its glass transition temperature, it shrinks into an equilibrium state, which gives rise to an approximately twofold decrease of its broad surface dimensions. Under these conditions the thin and stiffer metal films buckle and crack, to create wrinkled or flake-like structures, the structure of which is dependent on the hardness of the metal (Fig. 3.7). The resulting structures can be characterized in a predictive way by wavelength (Efimenko et al. 2005):

$$\lambda_{\omega} = 2\pi h_s \sqrt[3]{E_S / E_B}, \tag{3.3}$$

where h_S is the thickness of the metal skin layer, E_S is Young's modulus of the metal skin, and E_B is the Young's modulus of the substrate (Efimenko et al. 2005). Such wrinkled structures were shown to perform well as surface enhanced fluorescence substrates (Freschauf et al. 2012), and were also reported, under certain conditions, to be bactericidal, hydrophobic and self-cleaning (Sharma et al. 2014).

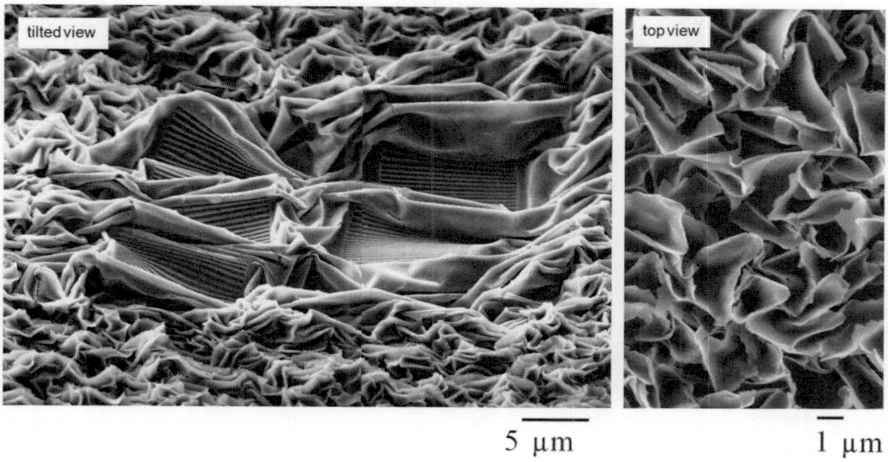

Fig. 3.7 SEM micrographs of a buckled 15 nm chromium layers on thermally shrunk plastic. Gratings of 400 nm in period were milled into chromium films before thermal rescaling. Ions implanted during milling locally altered the mechanical properties which govern buckling (Eq. 3.3)

3.4 Discussion and Conclusions

The common thread within all of the fabrication approaches discussed above is their parallel processing feature, which makes fabrication of antibacterial surfaces both simple and efficient. In the case of chemical etching, a wet bath process does not require the vacuum conditions necessary for the fabrication of black silicon. In this chapter, the possibility of using such nano-textured surfaces for sensing via SERS (Gervinskas et al. 2013b), for bactericidal applications (Ivanova et al. 2013), and for incorporation into micro-fluidic platforms, was discussed.

The photo-electrochemical etching of black silicon was demonstrated for the n-type silicon substrate. This step allowed the modification of the aspect ratio of the nanopillar on the surface of black silicon to be altered without changing the separation distance between the nanopillar. Usually, taller nanopillar with a higher aspect ratio are fabricated at the expense of larger inter-nanopillar separation (thus requiring a longer etching time). The aspect ratio, height, and inner-nanopillar separation are the key parameters for the recently observed bactericidal properties of black silicon (Ivanova et al. 2013). Hence, effective control methods for the independent adjustment of those parameters that are critical for the antibacterial behavior of the surface are highly desirable, with photo – electrochemical cell methods providing one of the possibilities for being able to do so.

The nano-textured surfaces of black silicon and black copper represent promising substrates for SERS detection processes after they have been coated with plasmonic metals, whereas buckled metal films on thermally shrunk plastic are efficient surface enhanced fluorescence substrates. SERS intensity is almost linearly proportional with the thickness of the deposited metal lm. For purposes of future streamlined SERS applications, incorporation of nano-textured electrodes inside microfluidic chips will allow the realization of new electrochemical or dielectrophoretic functionalities.

Acknowledgements We acknowledge G. Gervinskas, M. Pierrette, J. Juodkazytė for collaboration on the development and fabrication of the black silicon and black copper samples. The authors are grateful for support via the Australian Research Council DP130101205 Discovery project.

References

Buividas R, Stoddart PR, Juodkazis S (2012) Laser fabricated ripple substrates for surface-enhanced Raman scattering. Ann Phys 524(11):L5–L10

Buividas R, Fahim N, Juodkazytė J, Juodkazis S (2013) Novel method to determine the actual surface area of a laser-nanotextured sensor. Appl Phys A 14(1):169–175

Chou A, Jaatinen E, Buividas R, Seniutinas G, Juodkazis S, Izake EL, Fredericks PM (2012) SERS substrate for detection of explosives. Nanoscale 4(23):7419–7424

Efimenko K, Rackaitis M, Manias E, Vaziri A, Mahadevan L, Genzer J (2005) Nested self-similar wrinkling patterns in skins. Nat Mater 4:293–297

Freschauf LR, McLane J, Sharma H, Khine M (2012) Shrink-induced superhydrophobic and antibacterial surfaces in consumer plastics. PLoS One 7(8):e40987

Gervinskas G, Seniutinas G, Rosa L, Juodkazis S (2013a) Arrays of arbitrarily shaped nanoparticles: overlay-errorless direct ion write. Adv Opt Mater 1(6):456–459

Gervinskas G, Seniutinas G, Hartley JS, Kandasamy S, Stoddart PR, Juodkazis S (2013b) Surface-enhanced Raman scattering sensing on black silicon. Ann Phys 525(12):907–914

Hasan J, Webb HK, Truong VK, Watson GS, Watson JA, Tobin MJ, Gervinskas G, Juodkazis S, Wang JY, Crawford RJ, Ivanova EP (2012) Spatial variations and temporal metastability of the self-cleaning and superhydrophobic properties of damselfly wings. Langmuir 28(50):17404–17409

Ivanova EP, Hasan J, Webb HK, Gervinskas G, Juodkazis S, Truong VK, Wu AHF, Lamb RN, Baulin V, Watson GS, Watson JA, Mainwaring DE, Crawford RJ (2013) Bactericidal activity of nanostructured black silicon. Nat Commun 4:2838

Jayawardhana S, Rosa L, Buividas R, Stoddart PR, Juodkazis S (2012) Light enhancement in surface-enhanced Raman scattering at oblique incidence. Photonic Sens 2(3):283–288

Juodkazis S, Yamasaki K, Matsuo S, Misawa H (2004) Glass transition- assisted microstructuring in polystyrene. Appl Phys Lett 84(4):514–516

Juodkazis K, Juodkazytė J, Šebeka B, Savickaja I, Juodkazis S (2013) Photoelectrochemistry of silicon in HF solution. J Solid State Electrochem 17(8):2269–2276

Juodkazytė J, Šebeka B, Savickaja I, Selskis A, Jasulaitienė V, Kalinauskas P (2013) Evaluation of electrochemically active surface area of photosensitive copper oxide nanostructures with extremely high surface rough- ness. Electrochim Acta 109:109–115

Juodkazytė J, Sˇebeka B, Savickaja I, Jagminas A, Jasulaitiene˙ V, Sel- skis A, Kovger J, Mack P (2014) Study on copper oxide stability in photoelectrochemical cell composed of nanostructured TiO_2 and Cu_xO electrodes. Electrochim Acta 137:363–371

Kondo T, Juodkazis S, Misawa H (2005) Reduction of capillary force for high-aspect ratio nanofabrication. Appl Phys A 81(8):1583–1586

Lapierre F, Brunet P, Coffinier Y, Thomy V, Blossey R, Boukher- roub R (2010) Electrowetting and droplet impalement experiments on superhydrophobic multiscale structures. Faraday Discuss 146:125–139

Lapierre F, Brunet P, Coffinier Y, Thomy V, Blossey R, Boukher- roub R (2011a) High sensitive matrix-free mass spectrometry analysis of peptides using silicon nanowires-based digital microfluidic device. Lab Chip 11:1620–1628

Lapierre F, Piret G, Drobecq H, Melnyk O, Coffinier Y, Thomy V, Boukherroub R (2011b) EWOD driven cleaning of bioparticles on hydrophobic and superhydrophobic surfaces. Lab Chip 11:490–496

Lapierre F, Piret G, Drobecq H, Melnyk O, Coffinier Y, Thomy V, Boukherroub R (2011c) High sensitive matrix-free mass spectrometry analysis of peptides using silicon nanowires-based digital microfluidic device. Lab Chip 9:1620–1628

Lapierre F, Jonsson-Niedziolka M, Coffinier Y, Boukherroub R, Thomy V (2013) Droplet transport by electrowetting: lets get rough! Microfluid Nanofluid 15(3):327–336

Nishijima Y, Khurgin JB, Rosa L, Fujiwara H, Juodkazis S (2013) Randomization of gold nanobrick arrays: a tool for SERS enhancement. Opt Express 21(11):13502–13514

Nishijima Y, Hashimoto Y, Rosa L, Khurgin JB, Juodkazis S (2014a) SERS scaling rules. Appl Phys A 117:647–650

Nishijima Y, Hashimoto Y, Rosa L, Khurgin JB, Juodkazis S (2014b) Scaling rules of SERS intensity. Adv Opt Mater 2(4):382–388

Park K-C, Choi HJ, Chang C-H, Cohen RE, McKinley GH, Barbastathis G (2012) Nanotextured silica surfaces with robust superhydrophobicity and omnidirectional broadband supertransmissivity. ACS Nano 6(5):3789–3799

Seniutinas G, Rosa L, Gervinskas G, Brasselet E, Juodkazis S (2013) 3D nano-structures for laser nano-manipulation. Beilstein J Nanotechnol 4(1):534–541

Sharma H, Digman MA, Felsinger N, Gratton E, Khine M (2014) Enhanced emission of fluorophores on shrink-induced wrinkled composite structures. Opt Mater Express 4:753–763

Song M-J, Hwang SW, Whang D (2010) Non-enzymatic electrochemical CuO nanoflowers sensor for hydrogen peroxide detection. Talanta 80:1648–1652

Tobin MJ, Puskar L, Hasan J, Webb HK, Hirschmugl CJ, Nasse MJ, Gervinskas G, Juodkazis S, Watson GS, Watson JA, Crawford RJ, Ivanova EP (2013) High spatial resolution mapping of superhydrophobic cicada wing surface chemistry using infrared microspectroscopy and infrared imaging at two synchrotron beamlines. J Synchrotron Radiat 20(3):482–489

Žukauskas A, Malinauskas M, Kadys A, Gervinskas G, Seniutinas G, Kandasamy S, Juodkazis S (2013) Black silicon: substrate for laser 3D micro/nano-polymerization. Opt Express 21(6):6901–6909

Chapter 4
Electroactive Anti-microbial Surfaces

Paul J. Molino, Binbin Zhang, and Michael J. Higgins

Abstract Electroactive materials are becoming an increasingly important component of many electronic devices designed to interface with biological systems. While much of this work has been driven towards developing electrical stimulation protocols and novel electroactive materials to enhance interfacing with mammalian cells and tissues for therapeutic biomedical applications, electrically driven processes have been shown to be highly tailorable and effective at preventing microbial fouling of the electrode surface. In this chapter we review the range of electrical stimulation paradigms that have been investigated to deactivate and/or repel microbial organisms from electrode surfaces. The mechanisms through which electrical stimulation acts to kill bacterial cells will be discussed, and the application of new polymeric electroactive materials that offer great scope to modulate materials chemistry and fabrication processes to further enhance antimicrobial activity will be reviewed. Finally we look forward towards the innovations that will bring forth the next generation of electroactive antimicrobial materials that promise to provide solutions for a range of diverse applications.

Keywords Conductive electroactive materials • Electrical stimulation • Pyrrole • Aniline • Thiophene

P.J. Molino (✉) • M.J. Higgins
ARC Centre of Excellence for Electromaterials Science, University of Wollongong, Wollongong, NSW 2522, Australia

ARC Research Hub for Australian Steel Manufacturing, Intelligent Polymer Research Institute, AIIM Facility, Innovation Campus, University of Wollongong, Wollongong, NSW 2522, Australia
e-mail: pmolino@uow.edu.au

B. Zhang
ARC Centre of Excellence for Electromaterials Science, University of Wollongong, Wollongong, NSW 2522, Australia

E.P. Ivanova, R.J. Crawford (eds.), *Antibacterial Surfaces*,
DOI 10.1007/978-3-319-18594-1_4

4.1 Introduction

Conductive electroactive materials designed to interface with their environment are being developed to provide a suite of cutting edge technologies that promise to revolutionise numerous industries. For example, electrodes and electroactive materials are already finding uses in biomedical applications. Metallic electrodes, composed generally of platinum, iridium or titanium nitride, are being used to deliver electrical stimuli directly to the cochlea in the inner ear to provide a sense of sound for the profoundly deaf, for vagus nerve stimulation to control epilepsy or Parkinson's disease, and in pacemakers to regulate cardiac rhythm (Clark and Hallworth 1976; Norlin et al. 2005; Rose and Robblee 1990). Electrochemical sensors are highly effective tools for monitoring chemical composition and process control in both liquids and gas for a range of industries (e.g. food preparation, environmental and biomedical), providing real-time remote monitoring without the need for sampling. Electrodes are also being continually developed for a suite of industries in the energy sector, including water splitting, energy storage and conversion. For all of the above applications, charge must be transmitted between the electrode and the interfacial environment where a range of electrochemical processes take place. These may be in the form of sensing an analyte or microbial organism through the measurement of electrochemical processes at the electrode surface (Ronkainen et al. 2010), transferring charge from the electrode to excitable tissues such as nerves to evoke nerve action potentials and/or cellular developmental and behavioural responses (Balint et al. 2012; Merrill et al. 2005), or converting liquid water into its constituent components (hydrogen and oxygen) through electrochemically driven water splitting processes (Doyle et al. 2013).

It has been known since the late 1700s that electrical stimuli can be used to interface and communicate with biological systems thanks to the seminal studies of Luigi Galvani. He demonstrated the capacity to generate movement in the legs of deceased frogs using an electrical stimuli (Galvani and Aldini 1792; Piccolino 1998). While this early work would spawn what would become the field of electrophysiology, it also highlighted our ability to interact directly with biological systems using electrical stimulation. It is thus no surprise that researchers eventually turned to probing the ability to use electrochemically driven control processes to prevent the development of biological fouling layers on surfaces, and as a method for microbial sterilisation.

Exploiting electrochemical processes for the inactivation, or killing, of microbial organisms has been investigated for several decades. Early work in this area focused on the efficacy of pulsed electric fields (PEFs) to kill a range of microbial organisms in an electrolyte suspension (Sale and Hamilton 1967; Hülsheger et al. 1981, 1983). PEF treatment involves the insertion of two electrodes in an electrolyte solution containing the relevant microbial organism, with a high potential (kV range) pulsed across the two electrodes. Electrical stimulation conditions of up to 20 kV/cm^2 and pulse durations in the μs range have demonstrated greater than 99 % lethal action against bacteria and yeast cells suspended in solution (Hülsheger et al. 1981, 1983;

Sale and Hamilton 1967; Grahl and Märkl 1996). Electric field strength and the pulse duration both significantly impacted the lethal efficiency of the treatment, with cell mortality determined to result from electroporation of the microbial cell wall, leading to increased permeability, cell inactivation and death (Hülsheger et al. 1983). PEF treatment has been shown to be effective against a range of microbial organisms, including bacteria (Hülsheger et al. 1981, 1983; Grahl and Märkl 1996; Beveridge et al. 2005; Pervez et al. 2013), yeasts (Hülsheger et al. 1983; Grahl and Märkl 1996) and algae (Satirapipathkul et al. 2008; Zhou et al. 2013), making this a potentially viable technique for a diverse range of industries and applications. In recent years it has been touted as a promising method to pasteurize foods at room temperature without modifying the food quality and taste (Barbosa-Cánovas et al. 2001), and has been investigated for use in industries including fruit juice production (Saldaña et al. 2011) and wine making (Puértolas et al. 2009, 2010).

PEF treatment is a particularly mild-treatment for the inactivation of microbial organisms in suspension, compared to other treatments such as heat treatment or the use of toxic compounds such as chlorine, that is realising new opportunities in a number of niche industries. However the high potentials required make this technique are only applicable for use under highly controlled conditions, and therefore preclude its safe application to a number of biologically sensitive systems, including most biomedical (*in vitro* and *in vivo*) and environmental applications. As such, the ability to use alternative electrical stimuli protocols and processes delivered from conductive substrates and materials to inhibit the development of microbial fouling layers specifically on surfaces has attracted considerable interest in the literature. These technologies have the potential to replace, or enhance, existing technologies, or in some cases provide solutions to microbial fouling of surfaces and structures that cannot be addressed efficiently using currently available methodologies.

This chapter aims to provide an insight into the work that has been performed in developing low-fouling conductive surfaces, and their potential development and application in a variety of industries. We will probe the observed effect of different modes of electrical stimulation, including constant current/potential, as well as an alternating current/potential, on adhered microbes on the electrode surface. The use of different conductive materials, including metals, electroactive composites, and organic conductors such as conducting polymers and graphene, will be discussed, along with ways by which these materials are being adapted to provide specific functionality for desired applications. Finally, we glance into the future to forecast where research in electroactive anti-microbial surfaces is likely to be heading, providing insights into the next wave of promising technologies that may emanate from this research field.

4.2 Electrical Stimulation Paradigms to Prevent Biofouling on Electroactive Substrates

4.2.1 Constant Electrical Stimulation

Polarising an electroactive surface with either a positive or negative charge represents the earliest approach to employing electrical stimuli to prevent microbial fouling directly on electroactive surfaces. Initial work in this area focused on the application of a constant electrical charge, either by controlling the potential or the current, supplied to the electrode. These studies identified that both current density and the choice of electrode material play an important role in guiding the inhibition of microbial adhesion to the electrode surface and the subsequent cell mortality. In some of the earliest work, silver, platinum, gold, stainless steel and copper electrodes were tested against *Escherichia coli* and *Staphylococcus aureus* cultured in agar medium (Spadaro et al. 1974). Constant currents in the range of 0.02–20 μA/mm^2 were used, with currents in the higher range resulting either in severe corrosion of the metal and/or the generation of deleterious electrolytic products. At lower potentials (0.4–4 μA/mm^2), where no changes in pH, gas production or corrosion were observed, the silver electrode demonstrated the best antimicrobial performance, not only preventing bacterial adhesion, but also generating a zone of inhibition around the electrode. The primary mechanism of action was determined to be iontophoresis, with a positive current generating the release of silver ions from the electrode surface at a concentration that was sufficiently lethal to bacterial cells resident near the electrode interface to render them inactive. This mechanism has been shown to be operative with other metal types, such as gold and platinum (Davis et al. 1991), however the ability for non-metallic electrodes, such as carbon (Davis et al. 1991), to also present antimicrobial activity under the same electrically stimulated conditions revealed the likelihood that other mechanisms were also at play.

The application of constant negative currents has also been shown to exhibit a significant repulsive effect on bacterial cells on or near the electrode surface. This perhaps is not surprising given that most bacteria present an overall negative charge at their physiological pH (Jucker et al. 1996), and thus this would result in electrostatic repulsion between the bacterial cell wall and the electrode surface. Several studies have investigated the use of negative currents to prevent bacterial colonisation of the inner surface of cannulae used in a range of medical procedures, with cannula-related infections making up a significant proportion of all hospital related infections (Spelman 2002). Carbon impregnated polyurethane (15 % loading) was employed to provide a conductive inner surface on a cannula, which when charged with a constant negative electric current, to repel bacteria from the electrode surface (Crocker et al. 1992; Liu et al. 1993). A constant negative current of −10 μA was sufficient to cause the migration of bacterial cells away from the cannulae surface when immersed in an agar plate supporting a lawn of bacterial cells within minutes, with all bacteria (*Staphylococcus epidermidis, S. aureus, Enterococcus faecalis, Proteus mirabilis, Pseudomonas aeruginosa and Candida albicans*) migrating at

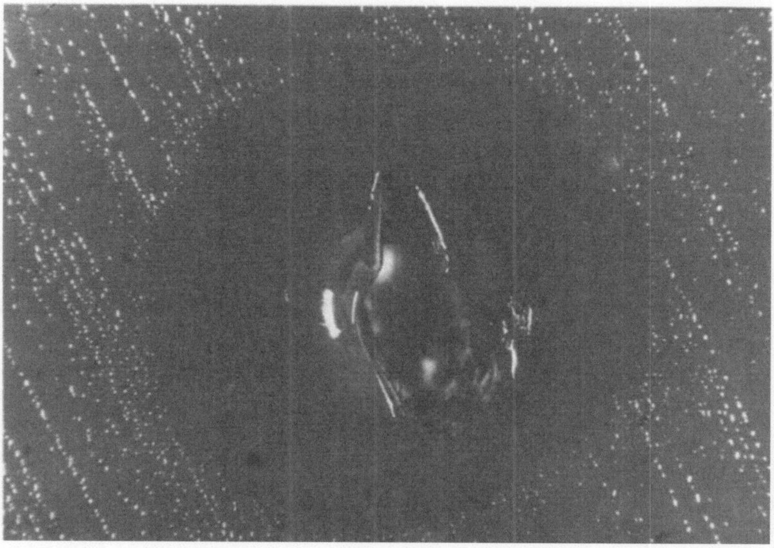

Fig. 4.1 Light micrograph of a nutrient agar plate seeded with *Staphylococcus epidermidis* showing a well-defined zone of inhibition around the site of a current carrying cannula. The cannula has been removed to facilitate demonstration of the zone (10× magnification) (Reproduced with permission from (Crocker et al. 1992))

least 100 μm from the cannulae surface within 15 min of the current being applied (Crocker et al. 1992) (Fig. 4.1). Varying the current strength directly influenced the size of the zone of inhibition, with −2, −10 and −50 μA generating zones of 4.0 ± 1.2 mm, 10 ± 2.4 mm and 14 ± 2.8 mm for *C. albicans*, respectively. The well-defined regions of inhibition surrounding the negative current carrying cannulae were determined to result from electrophoretic action against the microbial cells, protecting the cannulae from colonisation by the bacteria. In a following study, a −10 μA current applied to the carbon impregnated cannulae generated a zone of inhibition when the negatively charged end was placed into agar supporting a bacterial lawn (Liu et al. 1993), however this effect was not demonstrated for cannulae charged with a positive current. They found that bacteria were more resilient to the effects of the applied negative current if allowed to incubate around the cannulae prior to the application of the electrical stimuli. Furthermore, with currents of −10 μA having shown no obvious side effects when delivered directly into the human heart (Liu et al. 1993), these electrical stimulation paradigms were deemed suitable for *in vivo* applications.

Although a majority of the work on applying constant electrical charge to prevent microbial fouling has used electrical current, constant potential has similarly been demonstrated to be effective at deactivating and/or repelling bacteria from the electrode surface. A modest imposed surface potential of −66 mV (versus a saturated colomel electrode (SCE)) on an indium tin oxide (ITO) electrode resulted in an 88 % reduction in bacterial fouling on the electrode from a heterogeneous marine bacterial population in seawater (Kerr et al. 1998). The negatively charged electrode was found to both repel bacteria, and kill cells that remained within the vicinity of the electrode surface. In another study, 10 day old biofilms of *P. aeruginosa* showed partial removal of the biofilm on stainless steel on the application of −1.5 V versus Ag|AgCl (Dargahi et al. 2014). The partial removal of the biofilm, during a 1 min application of the imposed potential, was driven by electrostatic repulsion between the biofilm and the electrode. At potentials negative of −1.5 V, biofilms were removed from the electrode surface within seconds, however this was attributed to the vigorous evolution of hydrogen gas at the electrode surface under these stimulation conditions, resulting in the violent mechanical removal of the biofilm from the substrate.

The strength of the applied negative potential has been correlated with the removal of cells adhered at the electrode surface (Poortinga et al. 2001). Therein it was shown that while initial cell adhesion to an ITO electrode was independent of the potential applied during cell adhesion to the electrode (−0.2, 0.1, and 0.5 V versus Ag|AgCl), applying a more negative potential post-adhesion resulted in an increase in desorption of cells from the electrode surface. One bacterial strain deviated from this trend, which was attributed to the presence of long cellular appendages on the cell surface that were able to penetrate the repulsive energy barrier towards adhesion at the electrode-solution interface, allowing it to adhere while the other cells were more easily removed.

4.2.2 Alternating Electrical Stimulation

Alternating the polarity of charge delivered to an electrode surface between positive and negative has been developed as a method for killing and releasing microbial fouling organisms on an electroactive substrate. Employing alternating potentials as a method to kill and release microbial organisms from the electrode surface has been tested as a method to reduce micro- and macro-fouling in the marine environment (Nakayama et al. 1998a, b; Matsunaga et al. 1998; Okochi et al. 1998). The marine bacterium *Vibrio alginolyticus* was allowed to adhere to an ITO electrode, and was first subjected to a positive potential for 30 min, followed by exposure to a negative potential for 10 min (Okochi et al. 1998). A positive potential of 1.1 V versus SCE was shown to kill 70 % of cells adhered to the electrode surface, without producing changes in pH or generation of chlorine. Subsequent application of a negative potential of −0.43 V resulted in 73 % desorption of cells resident on the electrode. In a separate study, a nylon plate and polyethylene terephthalate (PET) film were coated with a conductive paint composed of graphite and carbon black mixed with urethane, and tested in the laboratory against fouling by *V. alginolyticus* (Matsunaga et al. 1998). Applying a potential of 1.2 V versus SCE for 30 min killed all cells adhered to the electrode surface, with a subsequent negative potential (−0.6 V) found to promote the subsequent removal of cells from the electrode interface. As observed in the previous study, the potentials investigated (−0.6 to 1.2 V) did not result in changes in pH or chlorine production. Field studies were conducted on the nylon netting coated in the conductive paint, with an alternating potential electrical stimulation protocol of 1.2 V versus Ag|AgCl for 60 min, followed by −0.6 V versus Ag|AgCl for 10 min, applied each day. After 158 days, the wet weight on the control nets were 65.5 g, while the netting with the applied potentials resulted in only 3.5 g of fouling, demonstrating the efficacy of the electroactive system at preventing marine fouling.

The nature of the electrode material has also been proposed to be critical in influencing the efficacy of electrical stimulation to act upon adhered microbial fouling organisms. Titanium Nitride (TiN) electrodes were shown to electrochemically inactivate *V. alginolyticus* cells at a lower potential than that demonstrated on conductive urethane coatings, illustrating that 98.7 % of cells were deactivated, or killed, on the electrode surface at a potential of 0.8 V versus Ag|AgCl for 30 min (Nakayama et al. 1998a). The efficacy of using TiN as a substrate to prevent biofouling was demonstrated by using a radio-frequency arc spraying technique (RFAS) to deposit the metal electrode on a PET plate, with the application of 0.8 V for 30 min killing 95.5 % of adhered *V. alginolyticus* cells in sterile seawater (Nakayama et al. 1998b). Field tests over 209 days demonstrated an alternating potential (1.0 V versus Ag|AgCl for 60 min, followed by −0.6 V Ag|AgCl for 10 min per day) to reduce biofouling by 96.3 % on the TiN electrode exposed to the alternating potential (5.1 g) versus the untreated electrode (134.7 g) (Fig. 4.2).

Alternating the current, as opposed to potential, has similarly been demonstrated to an effective means to deactivate and remove cells from an electrode surface. One study compared the effectiveness of applying negative, positive, and alternating currents to inhibit the adhesion of the bacteria *P. aeruginosa* to an ITO electrode (Shim et al. 2011). Therein *P. aeruginosa* cells exposed to a negative current of 15 $\mu A/cm^2$ exhibited a reduction of ~81 %, while positive currents illustrated comparable cell numbers to that on the unstimulated controls. Alternating the current between ±15 $\mu A/cm^2$ with 1 min intervals for each stimulation condition resulted in a similar reduction in adhered bacteria to the negative current treatment, however additionally resulted in a bactericidal effect on the adhered bacteria, making it the most suitable of the three electrical stimulation conditions (Fig. 4.3).

An alternating current electrical stimulation protocol, versus a constant current paradigm, was tested against biofilms of *S. epidermis* adhered onto surgical stainless steel electrodes for 200 min (van der Borden et al. 2004). *S. epidermis* is a common biofilm-forming bacteria that is a major cause of infection for orthopaedic implants. Four currents were investigated, −60 and −100 μA of constant current, and −60 and −100 μA of alternating current (50 % duty cycle, 1 Hz), applied for 360 min. Constant currents were found to be the most effective for removing adhered bacterial cells from the electrode surface at both current levels, with alternating currents demonstrating 24 % and 31 %, and constant currents 37 % and 78 %, for −60 μA and −100 μA, respectively. In addition, bacteria remaining on the electrode surface after the electrical stimulation were found to be less viable than prior to the treatment (97 % viability without treatment, 3 % and 2 % after −100 μA constant current and −100 μA alternating current treatments, respectively).

Fig. 4.2 Polyethylene terephthalate plates coated in a TiN electrode after 209 days with (**a**) and without (**b**) application of an alternating potential of 1.0 V against Ag|AgCl for 60 min and −0.6 V against Ag|AgCl for 10 min. The experimental period was 9th July 1997 to 3rd February 1998 (Reproduced with permission from (Nakayama et al. 1998b))

While most studies employ currents in the μA range, there have been investigations that apply even smaller currents, in the nA scale, in order to gauge their ability to act on biological fouling at the electrode interface. Microcurrents have been tested to inhibit the development of a conditioning film from urinary deposits, which aid in bacterial adhesion (Gabi et al. 2011). The development of the conditioning film and bacterial adhesion on platinum, proposed as a possible surface coating for urological stents, was studied using atomic force microscopy and quartz crystal microbalance. Alternating currents of +75, +320, and +750 nA/cm^2 (4 s stimulation cycle) resulted in either minimal or transient adsorption of compounds from the artificial urine. While a constant positive current resulted in relatively high adsorption of compounds from the artificial urine, only marginal film formation was

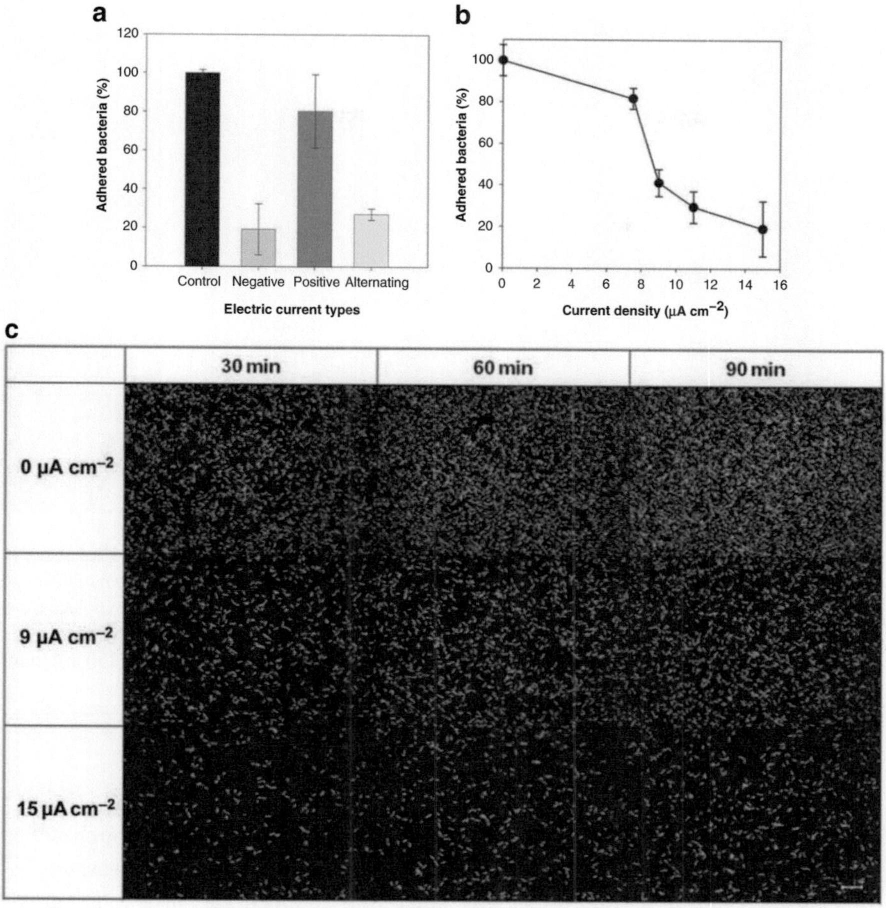

Fig. 4.3 (**a**) The numbers of adhered bacteria depending on the type of electric current (90 min, 15 μA.cm^2, $[KH_2PO_4]_0 = 20$ mM as base electrolyte, shear rate: 1.11 s^{-1}). Adhered bacteria were measured after switching off electric currents. $n = 3$; standard deviation shown; control: 100 % ± 1.7, negative current: 19.2 % ± 13.1, positive current: 80.4 % ± 18.9, alternating current: 27.0 % ± 2.9. (**b**, **c**) Bacterial adhesion during the application of the *negative current* depending on the electric current density and the time ($[KH_2PO_4]_0 = 20$ mM as base electrolyte, shear rate: 1.11 s^{-1}). (**b**) The numbers of adhered bacteria compared with non-polarization (at 90 min). n = 3; standard deviation shown; zero current: 100.0 % ± 7.5, 7.5 μA.cm^2: 81.8 % ± 5.0, 9.0 μA.cm^2: 41.1 % ± 6.5, 11 μA.cm^2: 29.5 % ± 7.6, 15 μA.cm^2: 19.2 % ± 3.1. (**c**) Images of adhered bacterial cells (*scale bar*=10 μm) (Reproduced with permission from (Shim et al. 2011))

observed upon the application of negative currents, even up to comparatively high current densities (−750 nA/cm^2). The application of mild alternating currents (75 and 320 nA/cm^2) was determined to reduce the bacterial colonisation of the platinum electrode surface by mediating the conditioning film formation, and alternating the microenvironment at the electrode interface (Fig. 4.4).

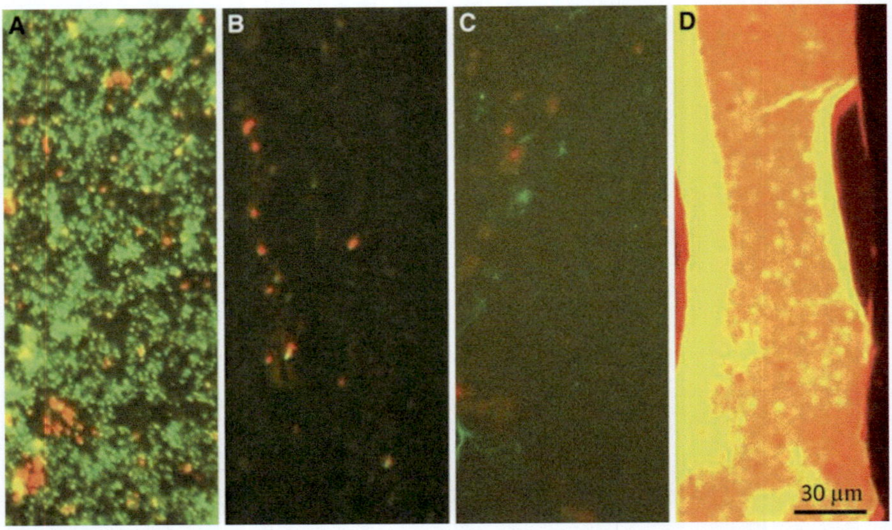

Fig. 4.4 Fluorescence microscopy images of four different study groups are shown after staining with fluorescein diacetate and propidium iodide to distinguish between viable (*green*) and dead (*red*) bacteria. Different current densities were applied to the platinum electrodes exposed for 6 days to artificial urine containing *Proteus mirabilis*. (**a**) No current $I=0$ applied. (**b**) alternating current $I=320$ nA/mm^2. (**c**) anodic current $I=75$ nA/mm^2. (**d**) anodic current $I=750$ nA/mm^2 (Reproduced with permission from (Gabi et al. 2011))

4.3 Redox Mediators

Several researchers have probed the principal mechanism of action through which electrical stimulation, either via current or potential, is able to deactivate or kill bacteria and other microbial organisms adhered to an electrode surface. Detection of bacterial cells using a cyclic voltammetry technique (CV) illustrated the ability to measure direct current transfer between cells from the bacteria *Saccharomyces cerevisiae* and a graphite electrode (Matsunaga and Namba 1984). The electron transfer peak at 0.74 V versus SCE was determined to be mediated by coenzyme A (CoA) which is associated with the bacterium cell wall. Subsequent work by this group found that the respiratory activity of *S. cerevisiae* decreased by up to 25 % when the electrode potential was held at 0.74 V (Matsunaga et al. 1984). The loss of respiratory activity was linked to the electrochemical oxidation of CoA in the bacterium cell wall, forming the dimeric CoA, resulting in the inhibition of respiratory activity leading to cell death.

Several research groups have investigated the use of redox mediators to improve the efficiency of electron transfer between the electrode and the microbial cell, allowing for a reduction in the potentials required to kill organisms, and therefore also reducing the likelihood of generating toxic substances such as hydrogen peroxide and chlorine (Okochi and Matsunaga 1997). Ferrocene is a fast and reversible redox mediator that is used in a number of industries, including those producing biosensors and organic solar cells (Yang et al. 2007; Daeneke et al. 2011). A graphite electrode modified via physisorption of ferrocene was found to deactivate and kill a greater percentage of adhered *V. alginolyticus* cells (0 % survival) at a far lower potential (0.2 V) compared to that of an unmodified electrode (5 % survival at 1.0 V) when treated for 30 min (Okochi and Matsunaga 1997). The presence of the ferrocene improved the electrochemical reaction efficiency, with cells adhered to the electrode surface during CV analysis generating an increase in the anodic peak current, indicating the ferrocene was mediating electron transfer between the electrode and the cells – a process that resulted in the efficient sterilization of the marine bacteria. Further work by this group determined that the lethal activity of the electrical potential did not result from an increase in the permeability of the cell membrane, but rather the electrocatalytic oxidation of intracellular substances (Okochi et al. 2000).

4.4 Innovative Electroactive Materials

Inherently conducting polymers (ICPs) are an exciting class of materials that possess a highly tuneable chemistry through which a range of polymer properties, including chemistry, conductivity, porosity, morphology, and surface energy, may be modulated and tailored for specific applications. ICPs such as pyrrole, aniline, and thiophene have been a particular focus for researchers due to their high conductivity, and their ability to engage a range of polymer synthesis methods (Wallace et al. 2002). ICP synthesis can be performed either electrochemically directly on an electrode surface, or chemically, forming a polymer nano-dispersion that can be used to fabricate ICP coatings, or composite coatings by employing them as an additive to a primary polymer system. ICP synthesis is facilitated by the oxidation of the monomer unit, forming a positively charged conjugated polymer backbone. This positive charge is counter balanced by the inclusion of an anionic species, termed the *dopant* (Wallace et al. 2002). A range of variables may be tailored to alter the fundamental material physicochemical and electrochemical properties, including the choice of dopant, synthesis method (chemical, electrochemical or vapour phase), and specific synthesis conditions (constant current, constant potential, CV, polymerisation time, charge density), making ICPs a highly flexible and tuneable polymer platform material.

Over the past decade, ICPs have attracted enormous interest in the areas of biomaterials and biological interfacing due to their good biocompatibility, as well as their ability to perform a range of biologically relevant functions, including delivery of electrical stimuli to excitable cells, controlled drug delivery, and guided cell growth. In addition to their ability to deliver faradaic charge at the electrode surface as per standard metallic electrodes, ICPs can undergo changes in polymer redox state, which, given the appropriate ICP chemistry, can result in a dramatic modulation of the polymer physicochemical properties (Halldorsson et al. 2009, 2011; Teh et al. 2009). Given this exciting array of adaptable polymer properties, it is surprising that researchers have only in recent years investigated the potential of employing electroactive ICPs in antimicrobial technologies and coatings.

The ICP polyaniline (PANI) is an excellent candidate material for many antimicrobial applications given its demonstrated high environmental stability and excellent anticorrosive properties (Kulkarni et al. 1989; Ansari and Keivani 2006; Biallozor and Kupniewska 2005). PANI used either as a dispersion, or as an additive in a range of traditional and novel composite marine coating systems, has demonstrated strong antimicrobial activity. Chemically polymerised polyaniline dispersions and nanoparticles have displayed strong inherent antibacterial properties against a range of bacterial strains (Gizdavic-Nikolaidis et al. 2011, 2012; Prasad et al. 2012; Jotiram et al. 2012). In one study, the antibacterial properties of PANI nanofibers were tested by analysing their effect on bacteria cultured in agar containing different concentrations of the nanofibers (5, 10, 15, 20 and 30 μg/mL) (Jotiram et al. 2012) (Fig. 4.5). The neat nanofibers were compared with fibers containing the antibiotic drug mupirocin (0.05 μg in 1 μg PANI nanofibers). PANI nanofibers alone demonstrated good antibacterial properties, generating zones of inhibition of 5 mm (5 μg/mL), 5.67 mm (10 μg/mL), 6 mm (15 μg/mL), 6.3 mm (20 μg/mL) and 7.3 mm

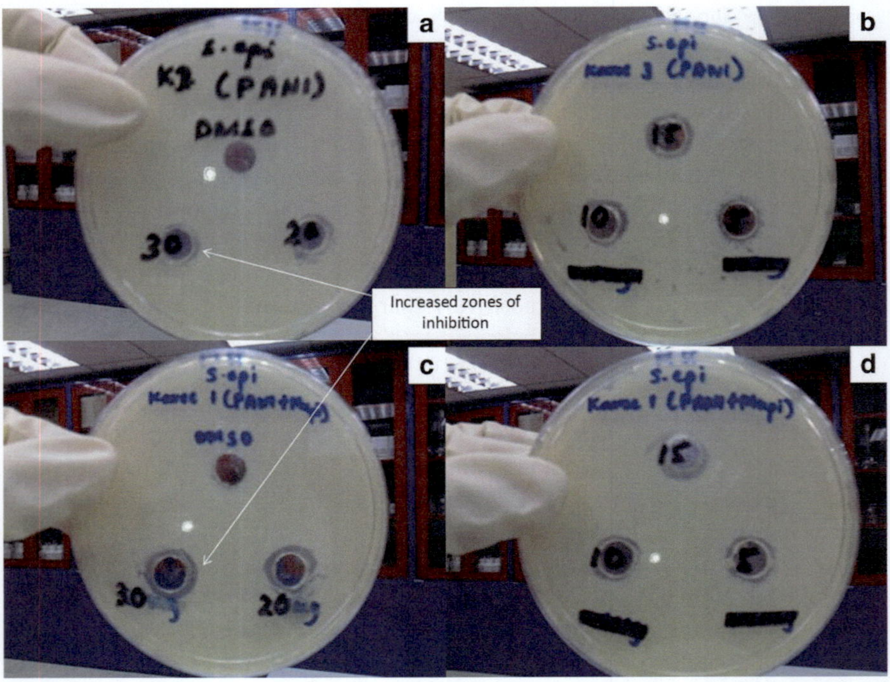

Fig. 4.5 Photographic image of zones of inhibition at various concentrations of PANI and PANI-mupirocin on *Staphylococcus epidermidis*. The antibacterial activity of PANI on *Staphylococcus epidermidis*: (**a**) Control dimethyl sulfoxide (DMSO) and PANI at 30 and 20 μg mL^{-1}. (**b**) PANI at 15, 10 and 5 μg mL^{-1}. (**c**) Control DMSO, PANI-mupirocin at 30 and 20 μg mL^{-1}. (**d**) PANI-mupirocin at 15, 10 and 5 μg mL^{-1} (Reproduced with permission from (Jotiram et al. 2012)).

(30 μg/mL) (Fig. 4.5). The antibacterial properties of the neat PANI were enhanced with the addition of mupirocin, increasing the zone of inhibition by between 6 and 21 %. A study into the mechanism/s behind the antimicrobial activity of PANI has determined that the polymer triggered oxidative stress in the bacterial cells, suppressing the formation of the bacterial cell wall (Gizdavic-Nikolaidis et al. 2011), providing an insight into the source of their potent antimicrobial properties.

Recently there has been growing interest in the application of PANI in marine surface coatings due to its ability to present both antimicrobial and anticorrosive properties (Wang et al. 1999; Mostafaei and Nasirpouri 2013; Yang et al. 2009). PANI and sulphonated PANI (sPANI) added to polyurethane (PU) and epoxy resin (EP)-based coatings were found to possess significant antifouling properties when the PANI comprised greater than 20 wt% of the coating (Wang et al. 1999). Additionally, the antifouling properties of the PANI were found to be closely related to its conductivity, with coatings possessing the doped conductive forms of the emeraldine salt demonstrating enhanced antifouling properties compared to the non-conductive dedoped forms. Both PANI and sPANI also demonstrated a synergistic behaviour when combined with toxic compounds, with an overall enhancement of the efficacy of coatings when containing either cuprous oxide or dichlorodiphenyltrichloroethane (DDT). While cuprous oxide – EP based coatings were effective for 2–3 months, the addition of doped PANI increased the effective life span of the coating for up to 9–12 months. The mechanism of enhancement of the antimicrobial effect of the copper oxide with the addition of the PANI was unclear, however was suggested to result from the PANI additive providing a local environment in the coating of ~pH 4–5, presenting a weak acidic microenvironment that enhanced the redox process of cuprous (I) – cupric (II) ion, improving its antifouling behaviour.

Films composed of nanocomposite blends of PANI with and without zinc oxide (ZnO) nanorods were demonstrated to present excellent antifouling properties when tested against marine fouling in the field over a period of 9 months, with EP coatings containing 4.5 wt% of PANI resulting in a significant decrease in marine biofouling from algae and barnacles compared to surfaces containing a lower wt% PANI and control EP coatings (Mostafaei and Nasirpouri 2013). Coatings containing 4.5 wt% PANI along with ZnO nanorods (up to 2 wt%) provided additional antifouling activity, likely due to the photocatalytic properties of the ZnO generating antibacterial agents such as superoxide ions, hydro-peroxide radical and OH; compounds that oxidatively stress the bacterial cells. The antimicrobial properties of both the PANI based EP coatings were supported by laboratory studies with the bacteria *E. coli* and *S. epidermis*, with the PANI-EP coating significantly inhibiting the growth of both bacterial strains compared to the EP control.

While ICPs such as PANI have been demonstrated to possess inherent antifouling qualities, such properties can be easily complemented by the incorporation of additional surface chemical functionalization via the use of thiol chemistry (Molino et al. 2012, 2013; Bergman and Hanks 2000). Such surface modification may have significant applications in preventing the fouling of ICP based electrochemical sen-

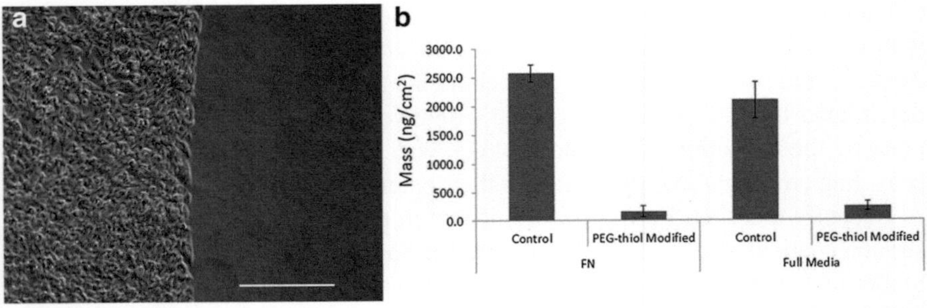

Fig. 4.6 (**a**) Colonisation of polyethylene glycol thiol (PEG-SH) modified polypyrrole-dodecyl sulfate (PPy-DS by primary mouse skeletal muscle cells (unmodified (*left*) and modified (*right*) polymer regions). (*Scale bar* represents 400 μm). (**b**) Mass of protein adsorption to PPy-DS polymer films with/without modification with PEG. PEG-SH modification undertaken with 40 k MW PEG at 1 mM concentration at pH 9 and 45 °C. Protein adsorption studied using Fibronectin (FN) (50 μg/mL in PBS) or full medium (HAMS F10 with 20 % foetal bovine serum). *Error bars* represent 95 % confidence intervals around the mean (Reproduced with permission from (Molino et al. 2013))

sors, preventing the fouling of stimulation and recording electrodes for biomedical applications, and providing enhanced antimicrobial properties to ICP materials engineered towards bestowing both anticorrosive and antimicrobial qualities to polymeric coatings. PANI and polypyrrole (PPy) have been shown to react readily with hydrophobic thiols to generate surfaces of dramatically decreased surface energy (Bergman and Hanks 2000; Molino et al. 2012). More recently, thiolated polymer brushes such as poly(ethylene glycol) have been employed to render the polymer surface resistant from protein and cellular interactions, providing a low fouling ICP electrode surface (Molino et al. 2013). PPy doped with the biological dopant dextran sulphate, an electroactive and biocompatible ICP that has exhibited excellent properties for bio-interfacing with cells and tissues, was functionalised to present a highly fouling resistant surface interface through the tethering of thiolated PEG (PEG-SH). Under optimal surface modification conditions, tethering of 40,000 Da PEG-SH reduced the nonspecific protein adsorption from HAMS F10 cell culture media containing 20 % foetal bovine serum (FBS) by 88 %, completely inhibiting the adhesion of mammalian skeletal muscle cells (Fig. 4.6). This approach is amenable to both the surface modification of ICP dispersions, as well as preformed films on a substrate, with the latter allowing the prospect of highly tuneable surface modification and patterning techniques such as inkjet printing of surface reactive chemistries such as PEG-SH, allowing control over the spatial distribution of low fouling chemistries on the ICP surface.

4.5 Conclusions and Future Directions

Electroactive materials have attracted significant interest in the field of antimicrobial coatings, where both electrical stimulation paradigms and electroactive materials chemistry are being explored to provide surfaces and devices that can be used to develop potent antimicrobial surfaces. While our ability to dissuade and impact microbial colonisation on metallic electrodes remains critical for many applications (particularly in the biomedical sector where choice of electrode is often restricted due to several factors), our ability to employ more flexible and tuneable conductors presents an enormous opportunity for their application in numerous industries. Organic conductors, such as carbon nanotubes and graphene, are being actively investigated as electrode materials for biological interfacing, and have already been shown to present inherent antifouling properties even without application of electrical stimuli (Hu et al. 2010; Kang et al. 2007). Further work on employing various electrical stimulation paradigms using these materials will be of great interest. As discussed, ICPs are an exciting class of materials that are highly processable and tuneable, and for which both electrically driven faradaic charge transfer, and modulation of polymer redox state to drive dynamic changes in polymer physicochemical and mechanical surface properties, may be exploited to act against the development of biofouling layers. Additionally, other electrochemically driven process may be utilised. For example, ICPs have been widely investigated for use in controlled drug delivery systems, where electrochemically controlled processes may be used to actively release drug compounds incorporated into the polymer. Such a process may be utilized to provide the controlled release of antimicrobial compounds from ICPs, which could be used individually or in concert with other mechanisms, to challenge microbial fouling organisms attempting to adhere to the electrode surface. Work by Esrafilzadeh and co-workers (Esrafilzadeh et al. 2013) have demonstrated the electrochemical release of the antibiotic drug ciprofloxacin from a PPy film in which the drug had been incorporated as the dopant species during electrochemical polymerisation. Application of a reducing potential to the polymer resulted in an excess negative charge on the polymer backbone, prompting the expulsion of the negatively charged ciprofloxacin from the polymer film and into the surrounding media. Ciprofloxacin release was shown to be effective against *E. coli* and *Streptococcus pyogenes*, generating a significant zone of inhibition around the stimulated ICP electrodes.

The specific development of electrically switchable surface chemistries may also provide promising opportunities for the generation of dynamic, electrically controllable antimicrobial surfaces and coatings. Switchable surface chemistries have been developed for use in biological systems, having been engineered to employ chemical, thermal, optical and electrochemical stimuli to transform the interfacial material properties (for review, see (Mendes 2008; Liu et al. 2005; Cole et al. 2009)). In particular, several studies have engineered surfaces that employ environmental stimuli, such as hydration, to trigger the reorganisation of surface chemistries to release microbial organisms such as bacteria from the material surface (Jiang and

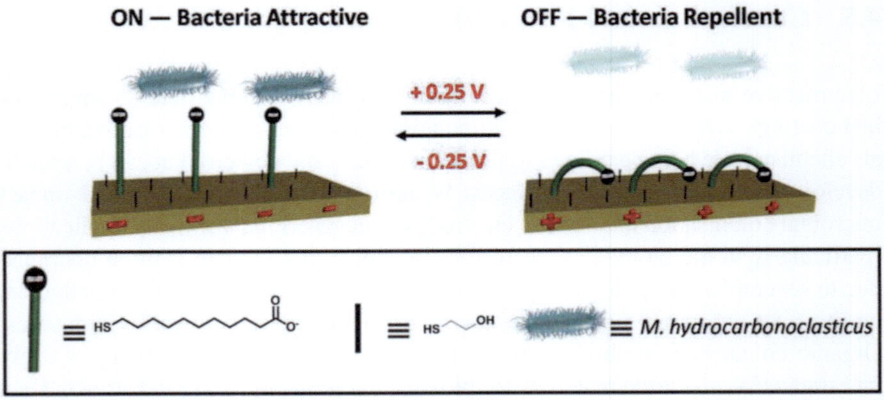

Fig. 4.7 Schematic representation of an electrically switchable two-component SAM that is able to reversibly and rapidly switch its molecular conformation in response to an applied potential. The change in molecular conformation induces either bacterial adhesion (anionic head group *exposed*) or repellence (anionic head group *concealed*) (Reproduced with permission from (Pranzetti et al. 2013))

Cao 2010). Surprisingly few studies have tested electrically switchable surfaces against microbial fouling organisms. Pranzetti et al. (2013) studied the early stages of bacterial cell adhesion to a switchable SAM surface that could modulate between attractive and repellent states. Their system was based on electrochemically switching the negatively charged 11-mercaptoundecanoic-acid (MUA), which was tethered to a gold surface along with a second thiol (mercaptoethanol (MET)) that acted as a spacer, by changing the polarity of the applied potential. Application of a negative potential exposed the anionic head group of the MUA, presenting an interface attractive to bacterial adhesion, while a positive potential resulted in the concealment of this head group towards the electrode surface, providing a bacterial repelling surface (Fig. 4.7). When in the bacterial repellent conformation, adhesion of the marine bacterium *Marinobacter hydrocarbonoclasticus* was reduced by ~83 % compared to surfaces switched to the attractive state. This study demonstrated the ability to exert fine electrical control over polymeric surface chemistries with antimicrobial activity, with the further development of surface chemistries that, rather than provide attractive – repulsive switchability, provide different or changeable antimicrobial properties (i.e. microbial killing (cationic ammonium quaternary system) and release (i.e. PEG, zwitterion functionality)) could provide promising intelligent surfaces for both the deactivation and release of microbes from the electrode coating surface. This approach forecasts the development of electrically switchable surface chemistries that provide fast and on-demand modulation of surface properties for effective release and/or killing of microbial organisms.

The testing and development of new electrical stimulation paradigms, coupled with the ongoing development of 'smart' and 'switchable' conductive electroactive materials and coatings may pave the way for the future development of electroactive materials that, in concert with the myriad of fabrication techniques that are under active investigation for these materials (i.e. wet spinning, electro-spinning, printing, knitting and braiding), provide flexible and highly tuneable materials that present innovative solutions applicable for a range of industries.

References

Ansari R, Keivani M (2006) Polyaniline conducting electroactive polymers thermal and environmental stability studies. J Chem 3(4):202–217

Balint R, Cassidy NJ, Cartmell SH (2012) Electrical stimulation: a novel tool for tissue engineering. Tissue Eng Part B Rev 19(1):48–57

Barbosa-Cánovas GV, Fernandez-Molina JJ, Swanson BG (2001) Pulsed electric fields: a novel technology for food preservation. Agron Food Ind Hi-Tech 12(2):9–14

Bergman B, Hanks TW (2000) Spectroscopic, microscopic, and surface analysis of alkanethiol- and fluoroalkanethiol-modified conducting polymer thin films. Macromolecules 33(21):8035–8042. doi:10.1021/ma000659p

Beveridge JR, MacGregor SJ, Anderson JG, Fouracre RA (2005) The influence of pulse duration on the inactivation of bacteria using monopolar and bipolar profile pulsed electric fields. IEEE Trans Plasma Sci 33(4):1287–1293. doi:10.1109/TPS.2005.852345

Biallozor S, Kupniewska A (2005) Conducting polymers electrodeposited on active metals. Synth Met 155(3):443–449. http://dx.doi.org/10.1016/j.synthmet.2005.09.002

Clark GM, Hallworth RJ (1976) A multiple-electrode array for a cochlear implant. J Laryngol Otol 90(07):623–627. doi:10.1017/S0022215100082529

Cole MA, Voelcker NH, Thissen H, Griesser HJ (2009) Stimuli-responsive interfaces and systems for the control of protein–surface and cell–surface interactions. Biomaterials 30(9):1827–1850. http://dx.doi.org/10.1016/j.biomaterials.2008.12.026

Crocker IC, Liu WK, Byrne PO, Elliott TSJ (1992) A novel electrical method for the prevention of microbial colonization of intravascular cannulae. J Hosp Infect 22(1):7–17. http://dx.doi.org/10.1016/0195-6701(92)90126-7

Daeneke T, Kwon T-H, Holmes AB, Duffy NW, Bach U, Spiccia L (2011) High-efficiency dye-sensitized solar cells with ferrocene-based electrolytes. Nat Chem 3(3):211–215. http://www.nature.com/nchem/journal/v3/n3/abs/nchem.966.html#supplementary-information

Dargahi M, Hosseinidoust Z, Tufenkji N, Omanovic S (2014) Investigating electrochemical removal of bacterial biofilms from stainless steel substrates. Colloids Surf B Biointerfaces 117(0):152–157. http://dx.doi.org/10.1016/j.colsurfb.2014.02.021

Davis CP, Wagle N, Anderson MD, Warren MM (1991) Bacterial and fungal killing by iontophoresis with long-lived electrodes. Antimicrob Agents Chemother 35(10):2131–2134. doi:10.1128/aac.35.10.2131

Doyle RL, Godwin IJ, Brandon MP, Lyons MEG (2013) Redox and electrochemical water splitting catalytic properties of hydrated metal oxide modified electrodes. Phys Chem Chem Phys 15(33):13737–13783. doi:10.1039/C3CP51213D

Esrafilzadeh D, Razal JM, Moulton SE, Stewart EM, Wallace GG (2013) Multifunctional conducting fibres with electrically controlled release of ciprofloxacin. J Control Release 169(3):313–320. http://dx.doi.org/10.1016/j.jconrel.2013.01.022

Gabi M, Hefermehl L, Lukic D, Zahn R, Vörös J, Eberli D (2011) Electrical microcurrent to prevent conditioning film and bacterial adhesion to urological stents. Urol Res 39(2):81–88. doi:10.1007/s00240-010-0284-3

Galvani L, Aldini G (1792) De Viribus Electricitatis In Motu Musculari Comentarius Cum Joannis Aldini Dissertatione Et Notis; Accesserunt Epistolae ad animalis electricitatis theoriam pertinentes. Apud Societatem Typographicam

Gizdavic-Nikolaidis MR, Bennett JR, Swift S, Easteal AJ, Ambrose M (2011) Broad spectrum antimicrobial activity of functionalized polyanilines. Acta Biomater 7(12):4204–4209. http://dx.doi.org/10.1016/j.actbio.2011.07.018

Gizdavic-Nikolaidis MR, Bennett J, Zujovic Z, Swift S, Bowmaker GA (2012) Characterization and antimicrobial efficacy of acetone extracted aniline oligomers. Synth Met 162(13–14):1114–1119. http://dx.doi.org/10.1016/j.synthmet.2012.04.031

Grahl T, Märkl H (1996) Killing of microorganisms by pulsed electric fields. Appl Microbiol Biotechnol 45(1–2):148–157. doi:10.1007/s002530050663

Halldorsson JA, Little SJ, Diamond D, Spinks G, Wallace G (2009) Controlled transport of droplets using conducting polymers. Langmuir 25(18):11137–11141. doi:10.1021/la900835w

Halldorsson JA, Wu Y, Brown HR, Spinks GM, Wallace GG (2011) Surfactant-controlled shape change of organic droplets using polypyrrole. Thin Solid Films 519(19):6486–6491

Hu W, Peng C, Luo W, Lv M, Li X, Li D, Huang Q, Fan C (2010) Graphene-based antibacterial paper. ACS Nano 4(7):4317–4323. doi:10.1021/nn101097v

Hülsheger H, Potel J, Niemann EG (1981) Killing of bacteria with electric pulses of high field strength. Radiat Environ Biophys 20(1):53–65. doi:10.1007/BF01323926

Hülsheger H, Potel J, Niemann EG (1983) Electric field effects on bacteria and yeast cells. Radiat Environ Biophys 22(2):149–162. doi:10.1007/BF01338893

Jiang S, Cao Z (2010) Ultralow-fouling, functionalizable, and hydrolyzable zwitterionic materials and their derivatives for biological applications. Adv Mater 22(9):920–932. doi:10.1002/adma.200901407

Jotiram KP, Prasad RGSV, Jakka VS, Aparna RSL, Phani AR (2012) Antibacterial activity of nanostructured polyaniline combined with mupirocin. Nano Biomed Eng 4(3): 144–149

Jucker BA, Harms H, Zehnder AJ (1996) Adhesion of the positively charged bacterium Stenotrophomonas (Xanthomonas) maltophilia 70401 to glass and Teflon. J Bacteriol 178(18):5472–5479

Kang S, Pinault M, Pfefferle LD, Elimelech M (2007) Single-walled carbon nanotubes exhibit strong antimicrobial activity. Langmuir 23(17):8670–8673. doi:10.1021/la701067r

Kerr A, Hodgkiess T, Cowling MJ, Beveridge CM, Smith MJ, Parr ACS (1998) A novel technique to prevent bacterial fouling, using imposed surface potential. J Appl Microbiol 85(6):1067–1072. doi:10.1111/j.1365-2672.1998.tb05272.x

Kulkarni VG, Campbell LD, Mathew WR (1989) Thermal stability of polyaniline. Synth Met 30(3):321–325. http://dx.doi.org/10.1016/0379-6779(89)90654-1

Liu W-K, Tebbs SE, Byrne PO, Elliott TSJ (1993) The effects of electric current on bacteria colonising intravenous catheters. J Infect 27(3):261–269. http://dx.doi.org/10.1016/0163-4453(93)92068-8

Liu Y, Mu L, Liu B, Kong J (2005) Controlled switchable surface. Chem A Eur J 11(9):2622–2631. doi:10.1002/chem.200400931

Matsunaga T, Namba Y (1984) Detection of microbial cells by cyclic voltammetry. Anal Chem 56(4):798–801. doi:10.1021/ac00268a047

Matsunaga T, Namba Y, Nakajima T (1984) 751—electrochemical sterilization of microbial cells. Bioelectrochem Bioenerg 13(4–6):393–400. http://dx.doi.org/10.1016/0302-4598(84)87040-3

Matsunaga T, Nakayama T, Wake H, Takahashi M, Okochi M, Nakamura N (1998) Prevention of marine biofouling using a conductive paint electrode. Biotechnol Bioeng 59(3):374–378. doi:10.1002/(SICI)1097-0290(19980805)59:3<374::AID-BIT14>3.0.CO;2-E

Mendes PM (2008) Stimuli-responsive surfaces for bio-applications. Chem Soc Rev 37(11):2512–2529. doi:10.1039/B714635N

Merrill DR, Bikson M, Jefferys JGR (2005) Electrical stimulation of excitable tissue: design of efficacious and safe protocols. J Neurosci Methods 141(2):171–198. http://dx.doi.org/10.1016/j.jneumeth.2004.10.020

Molino PJ, Wallace GG, Hanks TW (2012) Hydrophobic conducting polymer films from post deposition thiol exposure. Synth Met 162(15–16):1464–1470. http://dx.doi.org/10.1016/j.synthmet.2012.06.013

Molino PJ, Zhang B, Wallace GG, Hanks TW (2013) Surface modification of polypyrrole/biopolymer composites for controlled protein and cellular adhesion. Biofouling 29(10):1155–1167. doi:10.1080/08927014.2013.830110

Mostafaei A, Nasirpouri F (2013) Preparation and characterization of a novel conducting nanocomposite blended with epoxy coating for antifouling and antibacterial applications. J Coat Technol Res 10(5):679–694. doi:10.1007/s11998-013-9487-1

Nakayama T, Wake H, Ozawa K, Kodama H, Nakamura N, Matsunaga T (1998a) Use of a titanium nitride for electrochemical inactivation of marine bacteria. Environ Sci Technol 32(6):798–801. doi:10.1021/es970578h

Nakayama T, Wake H, Ozawa K, Nakamura N, Matsunaga T (1998b) Electrochemical prevention of marine biofouling on a novel titanium-nitride-coated plate formed by radio-frequency arc spraying. Appl Microbiol Biotechnol 50(4):502–508. doi:10.1007/s002530051327

Norlin A, Pan J, Leygraf C (2005) Investigation of electrochemical behavior of stimulation/sensing materials for pacemaker electrode applications: I. Pt, Ti, and TiN coated electrodes. J Electrochem Soc 152(2):J7–J15. doi:10.1149/1.1842092

Okochi M, Matsunaga T (1997) Electrochemical sterilization of bacteria using a graphite electrode modified with adsorbed ferrocene. Electrochim Acta 42(20–22):3247–3250. http://dx.doi.org/10.1016/S0013-4686(97)00174-6

Okochi M, Nakamura N, Matsunaga T (1998) Electrochemical control of bacterial cell accumulation on submerged glass surfaces. Clean Prod Process 1(1):53–59. doi:10.1007/s100980050008

Okochi M, Nakamura N, Matsunaga T (2000) Electrochemical killing of vibrio alginolyticus using ferrocene-modified electrode. Electrochim Acta 45(18):2917–2921. http://dx.doi.org/10.1016/S0013-4686(00)00368-6

Pervez N, Rasheed A, Pervez A, Rashid A, Mahmood Q (2013) Microbial inactivation utilizing impulse waveform. Desalination Water Treat 52(13–15):2662–2667. doi:10.1080/19443994.2013.794708

Piccolino M (1998) Animal electricity and the birth of electrophysiology: the legacy of Luigi Galvani. Brain Res Bull 46(5):381–407. http://dx.doi.org/10.1016/S0361-9230(98)00026-4

Poortinga AT, Bos R, Busscher HJ (2001) Reversibility of bacterial adhesion at an electrode surface. Langmuir 17(9):2851–2856. doi:10.1021/la001673y

Pranzetti A, Mieszkin S, Iqbal P, Rawson FJ, Callow ME, Callow JA, Koelsch P, Preece JA, Mendes PM (2013) An electrically reversible switchable surface to control and study early bacterial adhesion dynamics in real-time. Adv Mater 25(15):2181–2185. doi:10.1002/adma.201204880

Prasad RGSV, Chaitanya KSV, Tejoram M, Basavaraju D, Rao KN, Kumar RR, Sreenivasan S, Phani AR (2012) Antibacterial properties of nanofiber structured conducting polyaniline synthesized by cost effective wet chemical process. J Pharmacol Res 5(1):370–373

Puértolas E, López N, Condón S, Raso J, Álvarez I (2009) Pulsed electric fields inactivation of wine spoilage yeast and bacteria. Int J Food Microbiol 130(1):49–55. http://dx.doi.org/10.1016/j.ijfoodmicro.2008.12.035

Puértolas E, López N, Condón S, Álvarez I, Raso J (2010) Potential applications of PEF to improve red wine quality. Trends Food Sci Technol 21(5):247–255. http://dx.doi.org/10.1016/j.tifs.2010.02.002

Ronkainen NJ, Halsall HB, Heineman WR (2010) Electrochemical biosensors. Chem Soc Rev 39(5):1747–1763. doi:10.1039/B714449K

Rose TL, Robblee LS (1990) Electrical stimulation with Pt electrodes. VIII. Electrochemically safe charge injection limits with 0.2 ms pulses (neuronal application). IEEE Trans Biomed Eng 37(11):1118–1120. doi:10.1109/10.61038

Saldaña G, Puértolas E, Monfort S, Raso J, Álvarez I (2011) Defining treatment conditions for pulsed electric field pasteurization of apple juice. Int J Food Microbiol 151(1):29–35. http://dx.doi.org/10.1016/j.ijfoodmicro.2011.07.033

Sale AJH, Hamilton WA (1967) Effects of high electric fields on microorganisms: I. Killing of bacteria and yeasts. Biochim Biophys Acta (BBA) Gen Subj 148(3):781–788. http://dx.doi.org/10.1016/0304-4165(67)90052-9

Satirapipathkul C, Iwakabe K, Habaki H, Kawasaki J (2008) Inactivation of harmful dinoflagellate (Alexandrium catenella) in ballast water by electric treatment. Ann Microbiol 58(2):297–301. doi:10.1007/BF03175333

Shim S, Hong SH, Tak Y, Yoon J (2011) Prevention of Pseudomonas aeruginosa adhesion by electric currents. Biofouling 27(2):217–224. doi:10.1080/08927014.2011.554831

Spadaro JA, Berger TJ, Barranco SD, Chapin SE, Becker RO (1974) Antibacterial effects of silver electrodes with weak direct current. Antimicrob Agents Chemother 6(5):637–642. doi:10.1128/aac.6.5.637

Spelman DW (2002) 2: hospital-acquired infections. Med J Aust 176(6):286–295

Teh KS, Takahashi Y, Yao Z, Lu Y-W (2009) Influence of redox-induced restructuring of polypyrrole on its surface morphology and wettability. Sensors Actuators A: Phys 155(1):113–119. http://dx.doi.org/10.1016/j.sna.2009.07.006

van der Borden AJ, van der Werf H, van der Mei HC, Busscher HJ (2004) Electric current-induced detachment of Staphylococcus epidermidis biofilms from surgical stainless steel. Appl Environ Microbiol 70(11):6871–6874. doi:10.1128/aem.70.11.6871-6874.2004

Wallace GG, Teasdale PR, Spinks GM, Kane-Maguire LAP (2002) Conductive electroactive polymers: intelligent materials systems. CRC Press, Boca Raton

Wang XH, Li J, Zhang JY, Sun ZC, Yu L, Jing XB, Wang FS, Sun ZX, Ye ZJ (1999) Polyaniline as marine antifouling and corrosion-prevention agent. Synth Met 102(1–3):1377–1380. http://dx.doi.org/10.1016/S0379-6779(98)00384-1

Yang X, Lu Y, Ma Y, Liu Z, Du F, Chen Y (2007) DNA electrochemical sensor based on an adduct of single-walled carbon nanotubes and ferrocene. Biotechnol Lett 29(11):1775–1779

Yang SC, Brown R, Ramotowski T, Tucker W, Maranda L, Chena R, Shen M (2009) New anti fouling coatings based on conductive polymers. No. URITC Project No. 0001032

Zhou JF, Xu LP, Feng DL, Hu AB, Xie TF (2013) Study on inactivation of microalgae in ship ballast water by pulsed electric field and heat treatment. Adv Mater Res 610:3163–3166

Chapter 5
Manufacture of Chemically Modified Antibacterial Surfaces

Hitesh Pingle, Peng-Yuan Wang, Sally McArthur, and Peter Kingshott

Abstract It is well-known that the prevention of biofilm formation on medical implants is a highly desirable outcome for world-wide patient care. The lack of new antibiotic discoveries and the build-up of resistance towards existing antibiotics, particularly in biofilms, are proving a major global challenge to infection control and causing high mortality rates, which is exacerbated by the contamination of implant surfaces. The chemical modification of implants has proved to be very promising in preventing bacterial attachment over the past few decades, but despite the huge consequential reduction in bacterial attachment rates, no solution yet exists for preventing the attachment of bacteria and the subsequent formation of biofilms; it is generally recognised in the field of biomaterials surface science that it is possible to change the rate but not the fate of biofilm formation. This chapter provides an overview of the state of the art in the field of manufacture of chemically modified surfaces that are produced in an effort to minimise the formation of biofilms. A summary of the role of adsorbed biomolecules present in the environment surrounding implants is provided. This is an additional complicating factor, since we hypothesise that preventing biomolecular adsorption onto the surface of implants is needed to resolve the issue of implant-centred infection. Such an outcome therefore requires the development of novel surface modification strategies and perhaps more than one concept to improve the effectiveness of a coating. In this respect we summarise techniques such as the use of hydrophilic polymer layers and plasma polymers, self-cleaning surfaces, and combinations of controlled release of bactericidal molecules. In addition, we propose other novel antibacterial surface modification strategies. We also highlight that preventing the initial attachment of bacteria is a fundamental requirement in preventing biofilm formation, since a biofilm can form once a single bacterium has attached to a substrate surface.

Keywords Chemical modification • Polymerisation • Plasma polymer coatings • PEG/PEO

H. Pingle • P.-Y. Wang • S. McArthur • P. Kingshott (✉)
Department of Chemistry and Biotechnology, Faculty of Science, Engineering and Technology, Swinburne University of Technology, Hawthorn 3122, VIC, Australia
e-mail: pkingshott@swin.edu.au

E.P. Ivanova, R.J. Crawford (eds.), *Antibacterial Surfaces*,
DOI 10.1007/978-3-319-18594-1_5

5.1 Introduction

The use of biomedical devices and implants such as intravascular and urinary catheters, artificial hips and ventilators, has been rapidly rising with a view to improving the health of many patients. Unfortunately, at the same time, infection from biomedical implants during and after surgery is also rising; this has resulted in a major global problem. Whenever bacteria attach to surfaces they produce exopolysaccharides (EPS), which facilitate bacterial aggregation that subsequently rapidly starts the development of a sophisticated biofilm (defined as a group of microorganisms that stick to each other on a surface) (Sutherland 2001), that shows extreme resistance to antibiotics and biocides (Stoodley et al. 2002; Stewart and William Costerton 2001; Fey 2010; Percival et al. 2011; Majik and Parvatkar 2014; Vasilev et al. 2011; Ma et al. 2012). Researchers have shown that to eradicate bacteria in biofilms, a dose that is around 1,000 times more concentrated than that of conventional antibiotics used to kill planktonic bacteria is required (Smith 2005; Hetrick and Schoenfisch 2006). Once a biofilm forms, the implant needs to be replaced or treated via repeated surgery that can have serious consequences, financial burdens and even the death of the patient (Bazaka et al. 2012). It is estimated that in the USA alone the annual cost of antibacterialresistance is around 20 billion US dollars; more than 54,000 patients have been infected by catheter-related urinary tract infections in hospital settings (Frieden 2013; Smith 2005). In addition, in intensive care units, almost 16,000 patients are infected each year by catheter-related blood stream infections (Lyte et al. 2003). These are responsible for 87 % of the total blood stream infections (Richards et al. 2000). More than 13,000 deaths have been associated with urinary tract infections, and more than 30,000 deaths have been attributed to bloodstream infections, with more than 8,000 deaths related to surgical site infections even with the use of conventional antibiotics every year (Klevens et al. 2007). Health-care-associated infections are also a huge problem in developing countries (Allegranzi et al. 2011), where it is estimated that infection rates are two to three times higher than those of developed countries (Rosenthal 2011).

To overcome the problem of bacterial biofilm formation, it is necessary to prevent the initial attachment of bacteria to a substrate, thereby preventing the formation of the biofilm. A considerable amount of research has been published that has attempted to address this problem using surface modification strategies aimed at preventing bacterial attachment; this work is promising, but in many cases conflicting results are reported. These contradicting results are attributable to: (1) different testing procedures for bacterial attachment in different laboratories; (2) different strains of bacteria being used for the studies; (3) a lack of comprehensive surface analysis leading to incorrect assumptions being made about the effectiveness of the chemical modification of the surface; and (4) different animal models being used that lead to conflicting *in vitro* results.

The type of surface modification strategy that is employed is usually determined by the desired mode of action against the potentially adhering bacteria, the proposed physical-chemical mechanisms for preventing bacteria attachment, and/or the influence of the biomolecules that create the so-called "conditioning films" that facilitate bacterial attachment.

The most promising strategies to prevent bacterial attachment have employed different types of natural and synthetic polymers as surface modifying agents, aimed at reducing the attachment of common pathogenic bacteria such as *Staphylococcus epidermidis*, *Staphylococcus aureus*, *Pseudomonas aeruginosa* and *Escherichia coli*. These strategies include the fabrication of surfaces using non-ionic, hydrophilic and synthetic polymers such as poly (ethylene glycol) (PEG) (Chapman et al. 2001), poly(oxazolone) (Waschinski et al. 2005), and poly(acrylamide) (Fundeanu et al. 2008). The proposed antimicrobial mechanisms for such systems are based on the notion that end-attached polymer brushes provide a steric barrier against biomolecule adsorption and therefore the subsequent bacterial attachment to the underlying substrate. In this case, the polymer chain graft density and chain length or molecular weight are critical parameters that need to be optimised (Szleifer 1997; Kingshott and Griesser 1999) Poly(carboxybetaine) (Grigoras et al. 2012) and other zwitterionic polymers (Li et al. 2008a; Sin et al. 2014) have also displayed the ability to create substrates that possess enhanced anti-fouling properties, with a recent study of zwitterionic poly(sulfobetaine methacrylate) (pSBMA) grafted stainless steel showing effective reductions in bacterial attachment up to 6 h after initial exposure to the bacteria (Fig. 5.1) (Sin et al. 2014). The physical principle involved is that the modified surface mimics cell membrane surfaces that are naturally repellent to bacteria, with the water structure at such interfaces providing an energetically unfavourable surface for bacterial interactions.

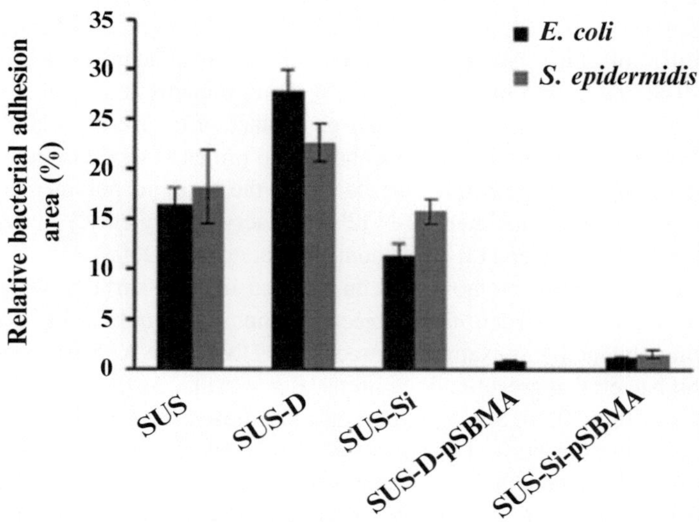

Fig. 5.1 The occupied area of *E. coli* and *S. epidermidis* on various surfaces after 6 h incubation. SUS: 316L type stainless steel coupons (SUS 316L). SUS-D and SUS-Si: dopamine- and silane-immobilized SUS surfaces, respectively. *pSBMA* poly(sulfobetaine methacrylate) (This chart was extracted from Sin et al. (2014) by permission from ACS Applied Materials and Interfaces)

Naturally derived polymers such as hyaluronic acid (Chua et al. 2008), alginic acid (Morra and Cassineli 1999) and surfactant type polymers (Vacheethasanee and Marchant 2000) have also shown promise at minimising biofilm formation, with the mechanism responsible being similar to that previously discussed. Bonilla et al. published a detailed review on polymers with antimicrobial properties (Muñoz-Bonilla and Fernández-García 2012). The building blocks of these polymers (e.g. ethylene glycol, carboxybetaine etc.) have also been used as the head groups of molecules that can self-assemble on surfaces. The classical alkane thiols on gold are one of the best examples, where the sulfhydryl group is attached to the gold substrate and the head group is exposed to the aqueous environment, the latter providing the antibacterial surface properties (Wiencek and Fletcher 1995; Hou et al. 2007). In light of all these promising attempts to prevent the attachment of bacteria, no surface yet exists that can achieve zero bacterial attachment. It is likely that the inability to identify an antibacterial surface derived from a molecular attachment process arises from the fact that it is difficult to obtain stable surfaces that are free of surface defects, together with the fact that there is still a lack of understanding regarding the function of these adsorbed molecules in the bacterial repellency process. It has been shown, however, that the covalent attachment of polymers onto substrate surfaces increases the stability of the film, which subsequently minimises the extent of bacterial attachment (Kingshott et al. 2003a).

Another promising approach for the prevention of bacterial attachment is through the use of naturally occurring bactericidal compounds such as essential oils (Kalemba and Kunicka 2003), plant products (Cowan 1999), and ions released from metallic nanoparticles embedded in the material. The effects of antibacterial furanones (Al-Bataineh et al. 2008) that potentially inhibit quorum sensing, or antimicrobial peptides (Costa et al. 2011) that penetrate cell walls to kill bacteria, are examples of molecules that are being used for the manufacture of antimicrobial coatings. The mode of function of such systems remains the subject of much research, but the ability to kill the bacteria on contact, or interfering with the bacterial adhesion processes are proposed as being the principal mode of action. Chen et al., for example, developed a surface using the cationic polymer poly(N,N-dimethyl-N-(ethoxycarbonylmethyl)-N-[2′-(methacryloyloxy)ethyl] ammonium bromide) that killed bacteria on initial contact (Cheng et al. 2008).

The role of adsorbed biomolecules on medical implant surfaces in restricting biofilm formation has been the subject of much research, with conflicting results being obtained. For example, extracellular DNA (eDNA) has been shown to both inhibit bacterial aggregation in biofilms (Berne et al. 2010) and facilitate adhesion (Tetz and Tetz 2010). Some proteins, such as those extracted from fish muscle and other food protein extracts, when adsorbed onto substrates have been shown to substantially reduce the number of bacteria adhering to the surface (Bernbom et al. 2006, 2009). Quite often, the methods needed to manufacture bacterial resistant coatings require multiple steps. For example, the attachment of polymer chains to medical polymers requires the use of a pre-treatment step where reactive groups are introduced onto the surface, which is followed by a chemical grafting or polymerisation step. One approach that has been exploited to create low fouling surfaces is

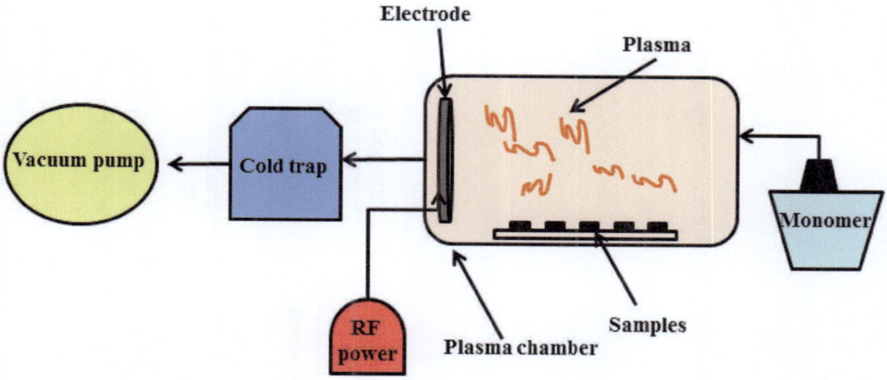

Fig. 5.2 Schematic illustration of a plasma reactor. Samples were placed in a vacuum chamber. By controlling flow rate or pressure of monomer in the chamber and radiofrequency (RF) power, the surface of biomaterials can be modified with a desired surface chemistry

plasma polymerisation (Fig. 5.2) (Triandafillu et al. 2003; Trentin et al. 2014; Ma et al. 2012). In this process, highly activated gas phase species react with inert surfaces generating thin films, where the chemical functionality of the film is determined by the choice of monomer used for the plasma polymerisation process. This has been used to either create low-fouling coatings in a single step using, for example, ethylene glycol type monomers, or to generate reactive species on surfaces for subsequent chemical attachment of non-fouling molecules (Jacobs et al. 2012).

Plasma polymer films have also been used for developing technologies that produce products that release antimicrobial agents in a controlled fashion, including the release of silver ions (Ag^+) (Vasilev et al. 2009). In this case, Ag nanoparticles are either loaded into or beneath a plasma polymer layer and the ion release kinetics are controlled by the concentration of particles in the film and/or the thickness of the film, which acts as a diffusion barrier. This concept is also applicable to other coatings, such as those generated by dip or spray coating or by impregnation of antimicrobial agents into the bulk of a material (Guo et al. 2013).

This chapter focuses on the strategies currently being used to control bacteria-surface interactions and the design of artificial, chemical-based antibacterial/antifouling surfaces that are relevant to medical implants. These include processes by which the chemistry of the surfaces is manipulated via polymer coatings or plasma polymer surfaces. The bacterial strains targeted and the role of biomolecules in the initial attachment of bacteria are also discussed. A summary of the different chemical modification strategies is provided in Table 5.1. Figure 5.3 depicts the typical chemical structures of poly(ethylene) glycol (PEG), poly (ethyleneimine) (PEI) (which is composed of repeating units of primary, secondary and tertiary amine groups), zwitterionic poly(sulfobetaine) methacrylate (pSBMA) (which contains both positively and negatively charged groups), the natural antibacterial polymer chitosan and the monomers allylamine and acrylic acid, both of which are used for plasma polymerisation.

Table 5.1 The different molecules/strategies used to chemically modify surfaces to prevent biofilm formation

Strategy	Molecules/materials	Surface modification method	Bacteria targete	Ref.
Hydrophilic synthetic polymers	Polyethylene glycol (PEG), poly(ethyleneimine) (PEI), poly(acrylic acid) (PAA)	Covalent attachment	*E. coli*, methicillin susceptible *S. aureus* (MSSA), methicillin resistant *S. aureus* (MRSA)	Pillai and Mutter (1981), Gudipati et al. (2005), and Ding et al. (2012)
Hydrophilic natural polymers	Chitosan, gelatin	solvent-casting method, cross linking with polymer using electron beam irradiation (EB)	*E. coli*, *S. aureus*, *S. typhimurium*, *P. aeruginosa*	Rhim et al. (2006), Zhao et al. (2003), and Jahani et al. (2014)
Zwitterionic polymers	Poly(sulfobetaine methacrylate) (pSBMA), poly(carboxybetaine methacrylate) (pCBMA)	Atom transfer radical polymerisation (ATRP)	*S. epidermis*, *P. aeruginosa*, *P. putida*, *E. coli*	Cheng et al. (2007), Li et al. (2008b), and Cheng et al. (2009)
				Li et al. (2008a) and Sin et al. (2014)

Self-assembled monolayers of non-fouling molecules	Oligo(ethylene glycol) (OEG), alkanethiolates	Self-assembly, ATRP	*E. coli*, *S. aureus*	Senaratne et al. (2005) and Deng et al. (1996)
Nanoparticles	Ag, Zn, Cu	Surface coatings, incorporation in the substrate material	*E. coli*, *S. aureus*, *B. subtilis*, *E. faecalis*, *S. pyogenes*	Kim et al. (2007), Ruparelia et al. (2008), and Jones et al. (2008)
Naturally derived small anti-fouling molecules	Cochliomycins A and B, sponge-derived polybrominated diphenyl ethers	Incubation in bacterial culture, coatings	*S. aureus*, *E. coli*, *Bacillus subtilis*, *Bacillus* sp., *Cytophaga* sp., *Zooshikella* sp. *Shewanella* sp.	Shao et al. (2011) and Ortlepp et al. (2008)
Antimicrobial peptides/enzymes/synthetic polymers	Cecropin, magainin, dermaseptin, subtilins, lysozyme, myeloperoxidase, polynorbornenes	Covalent bond, polymerization	*P. aeruginosa*, *S. aureus*, *E. coli*	Nguyen et al. (2011), Oren and Shai (1998), Zasloff (2002), Thallinger et al. (2013), Timofeeva and Kleshcheva (2011), and Al-Badri et al. (2008)

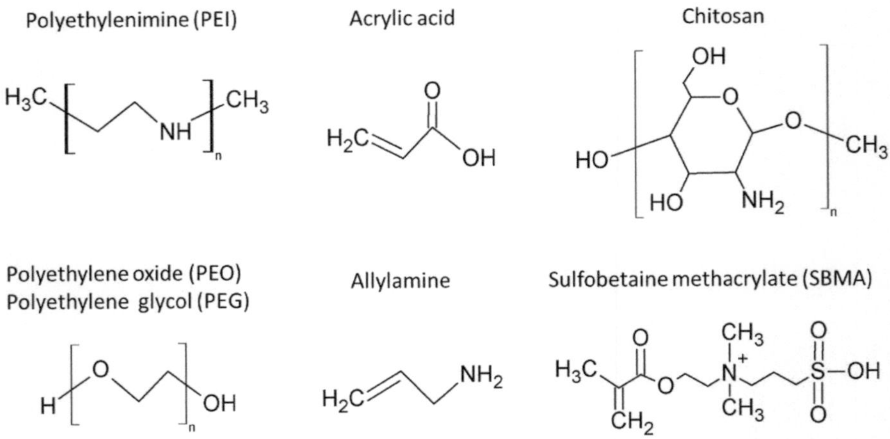

Fig. 5.3 Chemical structure of some common monomers (acrylic acid, allylamine, and sulfobetaine methacrylate (SBMA)) and polymers (PEI, PEO/PEG, and chitosan) used for surface treatment

5.2 Influence of Adsorbed Biomolecules on Biofilm Formation

Biomolecules such as polysaccharides, proteins, peptides, and lipids play an essential role in the initial attachment of bacteria to a substrate, which is usually followed by biofilm formation. Unless surface modification strategies are employed to prevent the adsorption of these biomolecules, they will spontaneously form so-called "conditioning films" that have been proposed to influence the extent of bacterial attachment and growth. Depending on the biological fluid (e.g. blood, urine, tears, interstitial fluid and bacterial growth media) to which the surfaces are exposed, the composition of the surface adsorbed bimolecular layer may change over time. This adds an additional level of complexity to the bacteria-surface interactions involved in the initial bacterial attachment. An open question in this area is whether the prevention of bacterial adsorption onto surfaces containing biomolecules is influenced by the surface alone, or whether other mechanisms also play a role (Poortinga et al. 2002). The precise way by which bacteria recognises eDNA (Whitchurch et al. 2002), protein (Heilmann et al. 1997) and/or carbohydrate (Kingshott et al. 2003b) components in conditioning films, however, remains elusive, however evidence is starting to emerge on this topic. Most pathogenic bacteria display cell surface adhesins, which have the tendency to recognise specific constituents present on the surface of mammalian cells. These adhesins also play a role in the initial bacterial attachment to the surface and initiate bacterial infection. There is a growing amount of evidence available that these adhesins may be associated with the recognition of biological constituents adsorbed onto surfaces of implanted devices, even though the specific mechanisms by which this occurs remain unexplained. For example, studies using *P. aeruginosa* have highlighted the role played by flagella and type IV pili in biofilm formation. Flagella motility is essential in allowing bacteria to make initial contact with abiotic surfaces and is also responsible for bringing the cells in close proximity to the surface, overcoming the repulsive forces that exist between the bacterium and the surfaces. Type IV pili also play a crucial role in stabilising the interaction between an abiotic surface and a bacterium, and mediating the twitching motility that is responsible for the migration of bacteria on surfaces (O'Toole and Kolter 1998). Waar et al. showed that surface proteins present on *E. faecalis* play an important role in controlling their adhesion to abiotic surfaces (Waar et al. 2002). *S. aureus* membrane surfaces also consist of proteins that are responsible for biofilm formation on abiotic surfaces (Cucarella et al. 2001). In addition, studies using *E. coli* have demonstrated that flagellar mediated motility is a key factor influencing biofilm formation under a range of conditions, including rich or poor media concentrations. Furthermore, surface associated proteins have been shown to be required for cell-cell interactions to occur during biofilm formation (Danese et al. 2000).

Recent research has shown that eDNA is a major component of biofilms and is also responsible for the maintenance and development of bacterial biofilms (Fey 2010; Gloag et al. 2013; Petersen et al. 2005). Many infectious bacterial species, including *P. aeruginosa* (Whitchurch et al. 2002; Gloag et al. 2013; Spoering and Gilmore 2006), *L. monocytogenes* (Harmsen et al. 2010), *B. cereus* (Vilain et al. 2009), *S. aureus* and *S. epidermidis* (Izano et al. 2008) have been implicated in the recognition of eDNA to enable attachment to a substrate and the consequent biofilm development on non-biological surfaces. A study using *Bordetella* biofilms showed that eDNA is one of the main components responsible for both the initial attachment of bacteria and the subsequent biofilm development. And the critical importance of DNase I to degrade biofilms formed inside an animal model of bacterial virulence (Fig. 5.4) (Conover et al. 2011). A study using *S. mutans* by Petersen et al. demonstrated the involvement of a synthetic competence-simulating peptide (SCAP) in the formation of a biofilm that involved the release of eDNA (Petersen et al. 2005), where the eDNA and a DNA binding uptake system played an important role in the mechanisms of biofilm formation. Another study by Conover et al. supported the postulation that eDNA plays an important role in maintaining *Bordetella bronchiseptica* (*B. brochiseptica*) biofilms in mice respiratory tracts, where the use of DNase I helped to degrade the biofilms (Conover et al. 2011). Such an approach may provide an insight in the treatment of bacterial infections in humans or in the development of bacterial resistant implant coatings. Despite these studies, the precise mechanisms by which eDNA promotes bacterial attachment and biofilm development is still unknown. Many of the cell surface components of *P. aeruginosa* that are involved in biofilm development have been identified. These include cup fimbriae (Vallet et al. 2004), type IV pili for irreversible attachment to surfaces (Barken et al. 2008), flagella to mediate initial reversible attachment (Yamamoto et al. 2012; O'Toole and Kolter 1998) and type IV pili to mediate surface migration (Klausen et al. 2003), micro-colony development and the formation of sophisticated biofilms (van Schaik et al. 2005). Type IV pili have also been identified as being able to bind to DNA (Barken et al. 2008), hence it is possible that these structures have been

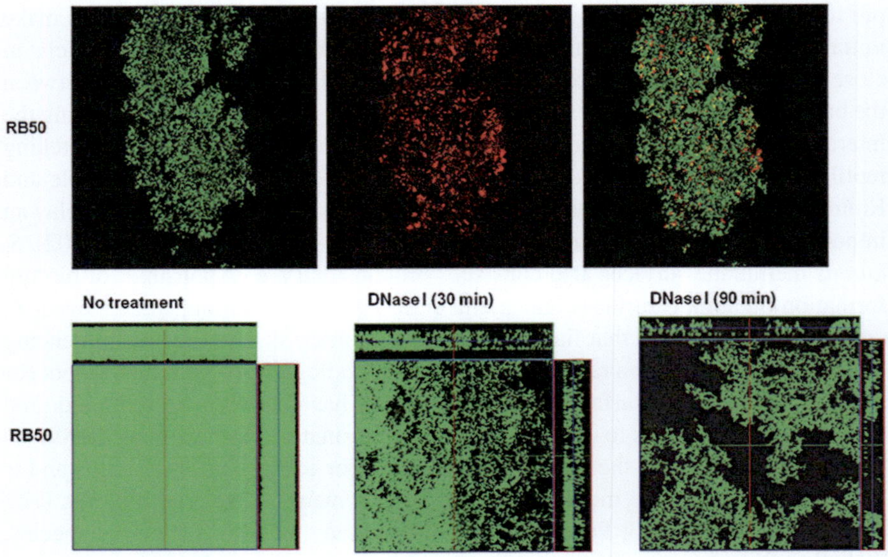

Fig. 5.4 The effect of eDNA on Bordetella biofilm. Three day old biofilms of B. bronchiseptica strain RB50 (*upper images*) stained with 7-hydroxy-9H-(1,3-dichloro-9,9-dimethylacridin-2-one) (DDAO). Confocal laser scanning microscopy (CLSM) images of live GFP expressing cells (*green*) and DDAO stained eDNA (*diffuse red*) or dead cells (*punctuate red*) are shown. *Yellow* appearance indicates the presence of both live cells and eDNA. DNase I lead to the disruption of established Bordetella biofilms grown on glass coverslips under static conditions (*lower images*). Biofilms were grown on glass coverslips for 48 h for RB50. The coverslips were gently rinsed followed by treatment with DNase I for either 30 or 90 min. For each micrograph, the *middle panel* represents the x-y plane, and the adjacent *top* and *side* panels represent the x–z and y–z planes, respectively. The image of a biofilm not treated with DNase I also depicted (The images were extracted from Conover et al. (2011) by permission from PLOS ONE)

involved in mediating bacterial attachment to surfaces through interaction with surface attached eDNA.

5.3 Biofilm Formation on Medical Implants

Percival et al. (2000) defined a biofilm as "microbial cells immobilised in a matrix of extracellular polymers acting as an independent functioning ecosystem, homeostatically regulated" (Percival et al. 2011; Percival et al. 2000). A biofilm is a convergence of microbial organisms that are irreversibly adhered onto surfaces, enclosed in a matrix of extracellular polymeric substances (Donlan and Costerton 2002). In terms of medical implants, infection via biofilm formation is caused by opportunistic pathogens such as *S. aureus* (Di Poto et al. 2009; Toté et al. 2009; Nablo et al. 2005), *S. epidermis* (Popat et al. 2007; Nablo et al. 2005), *P. aeruginosa* (Nablo et al. 2005; Toté et al. 2009) and *E. coli* (Carmen et al. 2005), on non-surgical medical devices such as intravenous and urinary catheters (Chenoweth and Saint 2013) and surgical medical devices such as prosthetic heart valves (Manne et al. 2012), cardiac pacemakers (Heimberger and Duma 1989), vascular prostheses (Moore et al. 1981), orthopaedic implants (Darouiche 2004), contact lenses (Dutta et al. 2012), intrauterine contraceptive devices (Castellsagué et al. 2011), artificial hearts (Griffith et al. 1988), voice prostheses (Post et al. 2004) and ocular prostheses (Zegans et al. 2002). There is a likelihood that non-surgical devices can become infected from pathogens from the skin at the time of insertion or after implantation via migration of pathogens onto the surfaces of these devices. Some other factors that may be responsible for surgical device infection are clot formation and tissue destruction at the implant site. In addition, adsorbed biological material tends to attract planktonic bacterial attachment and these factors increase the likelihood of biofilm development. It has been suggested that after successful implantation of a medical device, there is a link between the conditioning media present in the surrounding tissue and the extent of bacterial attachment onto implants (Rodrigues 2011). The first 6 h after implantation is the most critical, as has been showed by Poelstra et al. during which time, suppression of the activity of the pathogen is mandatory for the long term stability of implants (Poelstra et al. 2002; Hetrick and Schoenfisch 2006).

Once the implant surface has been conditioned via bio-fluids such as plasma proteins, lipids, extracellular matrix molecules, and inorganic salts, bacteria commence the reversible attachment with the surface. At a later stage, irreversible attachment to the surface occurs. This results in the secretion of insoluble gelatinous exopolymers, which allows the formation of a biofilm, which protects the bacteria from certain environmental threats such as biocides, antibodies, antibiotics, bacteriophage, surfactants and also from white blood cells and free flowing amoebae (Dunne 2002). Most recent research has shown that bacteria have a very high resistance generation capability against almost all antimicrobial agents when present in biofilms. Therefore, the best possible solution to overcome the protective nature of biofilms is to develop surfaces that can resist or reduce the initial attachment of microbes, preventing the subsequent formation of a biofilm.

5.4 Artificial Antibacterial Surfaces Based on the Chemical Modification of the Surface

In an effort to reduce the number of infections arising from implant devices, there has been a focus on understanding the surface chemistry of biomedical implant materials with a view to developing effective and long lasting antibacterial surfaces using traditional and highly advanced surface modification techniques (Hasan et al. 2013; Timofeeva and Kleshcheva 2011). These types of surfaces are made of different types of materials including anti-adhesive/antibacterial polymers and bioactive antibacterial molecules. Polymers, ceramics and bioactive ingredients have also been combined to generate the controlled release of antibacterial agents.

Nanostructured materials such as nanoparticles of Ag, Au, Cu, Zn and Ti have also been extensively used in the development of antimicrobial surfaces (Campoccia et al. 2013). The chemical surface treatments that have been used include surface functionalization and derivatisation/polymerisation techniques (Tiller et al. 2001; Hasan et al. 2013).

5.4.1 Polymerisation as a Way to Modify Surfaces

Conventional polymerisation is a highly versatile technique that can be exploited to modify the surface of a material so that it displays antimicrobial properties. Antimicrobial surfaces need to be stable, easy to synthesise, insoluble in aqueous medium, able to be regenerated and exhibit biocidal activity towards a broad spectrum of bacteria (Iarikov et al. 2013). Factors that influence the effectiveness of antimicrobial coatings include the surface charge density, polymer molecular weight, degree of alkylation and the hydrophilic/hydrophobic balance. Synthetic medical polymers such as nylon, poly(ethylene terephthalate), poly(fluoroethylene), polyethylene, and polypropylene have been surface modified by the covalent attachment of antimicrobial agents (Tiller et al. 2001; Lin et al. 2003; Hasan et al. 2013). Tiller et al. demonstrated (Tiller et al. 2001) that attached polymer chains could be polycationic, hydrophobic and should be sufficiently long to exhibit antimicrobial characteristics. One widely used surface polymerisation technique is covalent attachment and atom transfer radical polymerisation (ATRP) (Hasan et al. 2013; Lee et al. 2004; He et al. 2011). Polycation based hydrogels showed antibacterial properties by disrupting the bacterial cell wall, displaying potency against both Gram positive and negative bacteria (Li et al. 2011). Other types of cationic polymers that have been used for this purpose used include chitosan (Goy et al. 2009) and polyamine films (Kwon and Lu 2007). One such surface was generated by the layer-by-layer (LbL) self-assembly of the polyelectrolyte (Jiang et al. 2009) quaternary ammonium hydrophobic cation poly (vinyl-N-hexylpyridinium). These surfaces displayed strong antibacterial properties, killing 90–99 % of bacteria that came into contact with the surface (Lin et al. 2002). Lin et al. covalently immobilised N-alkylated poly(ethylenimines) (PEIs) onto textile materials, including cotton, wool, nylon and polyesters, to introduce antimicrobial properties to these fabrics (Lin et al. 2003). Their results showed that high molecular weight PEI chains exhibited higher degrees of antibacterial activity in comparison to low molecular weight PEIs, which showed around 98 % bactericidal activity. Some uncertainties were identified regarding the properties that were responsible for the antimicrobial activity; these included the proper control over monomer distribution, molecular weight, polydispersity and density of functional groups. To overcome this problem, atom transfer radical polymerisation (ATRP) was applied to quaternary ammonium groups by Lee et al. who showed greater control over the polydispersity and molecular weight of the resultant antimicrobial layer (Lee et al. 2004). The main advantages of these types of surfaces are the longer lasting antibacterial behaviour and the ease of synthesis of these films on common substrate materials, such as glass and paper. These types of surfaces can be used in food packaging, everyday household items and in military applications. Research is on-going on for wider scale applications (He et al. 2011), however, the possibility of application of such films to medical implants is doubtful due to their potential cytotoxicity.

Another approach towards surface modification includes the non-covalent physiochemical adsorption of antimicrobial agents such as antibacterial polymers, peptides, and enzymes (Hasan et al. 2013; Izadpanah and Gallo 2005; Timofeeva and Kleshcheva 2011). In 1965, Cornell and Dunraruma formed polymers and copolymers of 2-methacryloxytroponones, which were able to kill bacteria (Cornell and Donaruma 1965). Since that time, a large number of new antibacterial polymeric materials have been synthesized and research is still underway to understand the actual mechanisms responsible for the antimicrobial activity. Sidenbidel et al. categorised antimicrobial polymers into three types according to their function i.e. polymeric biocides, biocidal polymers and biocide releasing polymers (Siedenbiedel and Tiller 2012). Dizman et al. synthesized a norfloxacin-containing methacrylate monomer and PEG methyl ether methacrylate, which showed comparable antimicrobial activities against both *S. aureus* and *E. coli* (Dizman et al. 2004). In terms of biocidal activity, however, these polymers do not contain any repeating units of antimicrobial agents. The intact macromolecule behaves as an antibiotic agent. Some examples of these types of antimicrobial polymers include polycations with phosphonium (Kanazawa et al. 1993a, b), quaternary ammonium (Timofeeva and Kleshcheva 2011; Kawabata and Nishiguchi 1988) and tertiary sulphonium groups (Chen et al. 2000). The development of polymers that containing antimicrobial agents that can be released in the presence of microbes is a valid approach; Kenawy et al. provided a detailed review describing the applications and chemistry of such antimicrobial polymers (Kenawy et al. 2007).

Another class of antibacterial agents are the antimicrobial peptides (AMPs), which are fragments of cationic amphiphilic peptides present in the immune defence systems of every living organism. AMPs exhibit broad spectrum antimicrobial activity against bacteria, fungi, and have the ability to encapsulate viruses and parasites. In comparison to antibiotics that attack specific proteins present in bacteria, AMPs act on the bacterial membranes, and hence this is less likelihood resistance being developed through gene mutation. These qualities suggest that AMP is a viable alternative to antibiotics (Nguyen et al. 2011; Hancock and Sahl 2006). Magainin is an α-helical AMP derived from the African clawed frog, and is, to date, the most studied AMP, and Hancock (Hancock and Sahl 2006) and Nguyen (Nguyen et al. 2011) have provided a detailed review in this area.

Another class of effective biomolecules are antimicrobial enzymes that are able to control or restrict biofilm growth (Hasan et al. 2013; Thallinger et al. 2013). Antimicrobial enzymes are omnipresent in nature and play significant roles in the defence mechanisms of living organisms. They have the capability to kill the microorganism through interference with biofilm formation or growth and though catalysis reactions are able to produce antimicrobial compounds. Antibacterial enzymes that have been studied include proteolytic enzymes (subtilins, lysostaphin and bacteriophages lysin), polysaccharide degrading enzymes (lysozyme, amylases, dispersin B and alginate lyase), and oxidative enzymes (myeloperoxidase, cellobiose dehydrogenase, lactoperoxidase, glucose oxidase, horseradish peroxidase) (Fuglsang et al. 1995; Thallinger et al. 2013). Liquid formulations of antibiofilm and antimicrobial agents containing these enzymes are generally associated with surface cleaning products. One or more enzymes can be combined with other types of antibacterial agents and can also be incorporated or attached to the surface of substrates for the generation of antibacterial surfaces (Augustin et al. 2004). For example, subtilisins are the most extensively applied enzymes used by industry to control the formation of biofilms. It is a hydrolytic enzyme, which hydrolyses the bacterial adhesin proteins, i.e. those responsible for bacterial attachment onto the surface, hence preventing attachment (Leroy et al. 2008). Polysaccharide degrading enzymes such as lysozyme are frequently used by the healthcare and food industries to kill both Gram positive and negative bacteria. Oxidative enzymes, which are produced by the human body for defence against microbes, produce superoxide anions that further form hydrogen peroxide (H_2O_2) to kill bacteria (Thallinger et al. 2013). Incorporation of all these types of antimicrobial agents and strategies in the formation of an effective and long term stable surface is a possible solution for the prevention of biofilm formation, however further research is required.

5.4.2 *Plasma Polymer Coatings*

Many different surface modification strategies have been developed in an attempt to overcome biofilm formation on medical implants, however, it is important that the surface modification strategy does not affect the functional properties of the implant material, such as the visual transparency of contact lenses or the flexibility of vascular grafts (Vasilev et al. 2011). One method that is able to generate chemical groups on traditionally inert materials such as biomedical polymers, but have minimal influence upon the integrity of the material, is plasma polymerisation (Fig. 5.2). The method employs a power source (direct current, alternating current, radiofrequency or microwave) to excite gas phase molecules. The resultant plasma glow discharge of ions and radiation is able to introduce chemical groups onto a substrate surface, the resulting groups being dependent on the type of monomer gas being used. Typically, monomers such as allylamine, acrylic acid, methanol, acetaldehyde, acid chlorides, to name a few, are used to create reactive surfaces by the plasma polymerisation of thin films onto surfaces. It is also possible to plasma treat surfaces with monomers such as ammonia, where functional groups are introduced into the backbone of a polymer or to physically etch surfaces, for example, with a water plasma (Bilek and McKenzie 2010; Bazaka et al. 2011). The main advantages of plasma polymerised surface modification include: (1) the ability to change the surface chemistry of polymers without affecting the bulk properties; (2) the ability to control the chemical functionality and topography of the surface; and (3) the ability to synthesize physically stable surfaces. Each of these advantages makes the technique very suitable for the formation of biocompatible, antibacterial surfaces. For example, a polyterpenol thin film generated from terpinen-4-ol using plasma polymer coating has been shown to inhibit bacteria adhesion and decreasing the biofilm formation (Fig. 5.5) (Bazaka et al. 2010). Plasma polymer interlayers have been found to be useful in the generation of non-fouling fouling coatings by covalently grafting fouling-resistant polymers such as PEG (Kingshott et al. 2002), poly(lysine)-PEG copolymers (Harris et al. 2004), and poly(acrylamide) onto the surface (Fundeanu et al. 2008). Antibacterialmetallic nanoparticles such as Ag (Chen et al. 2008), Cu (Daniel et al. 2009) and ZnO (Applerot et al. 2010) have been incorporated into plasma polymer layers to enable the controlled release of ions that are capable of killing the bacteria (Garcia-Fernandez et al. 2012; Hasan et al. 2013). Plasma polymer layers also have been used as sinks for the deposition of antibiotics, also for controlled-release purposes. Kwok et al. used a n-butyl methacrylate plasma polymer as a diffusion barrier for ciprofloxacin loaded polyurethane, again for controlled release applications (Kwok et al. 1999). Overall, plasma coated surfaces show huge potential in the development of antibacterial and drug release systems, with the choice of monomers that can be applied being almost limitless.

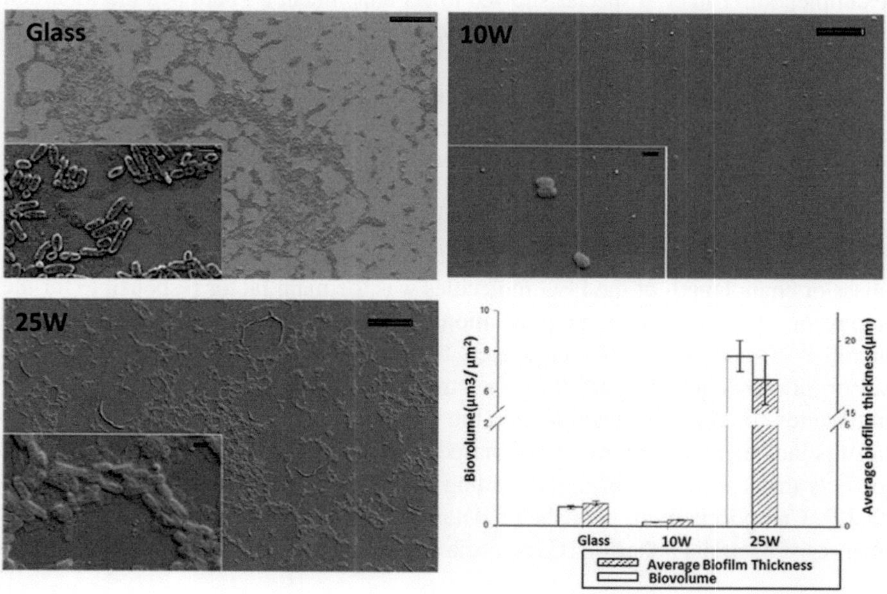

Fig. 5.5 *P. aeruginosa* on different surfaces after 18 h incubation. Scanning electron microscopy (SEM) images of *P. aeruginosa* on the glass (*top left*) and polyterpenol thin films fabricated at 10 W (*top right*) and 25 W (*bottom left*) radiofrequency (RF) deposition power represent an overview of the attachment patterns taken at 500× magnification (*bars* = 20 μm); *insets* are zoomed in areas taken at a 5000× magnification (bars = 1 μm). For (*bottom right*) each sample, at least four random locations of approximately 126.7 × 126.7 μm areas were scanned. Quantification of the *P. aeruginosa* biovolume and average biofilm thickness on the same surfaces indicated considerable inhibition of bacterial growth on the polyterpenol thin films fabricated at 10 W RF deposition power (These images were extracted from Bazaka et al. (2010) by permission from ACS Biomacromolecules)

5.4.3 PEG/PEO Based Approaches

Immobilising poly(ethylene glycol) (PEG; M.W. < 100,000) or poly(ethylene oxide) (PEO; M.W. >100,000) on surfaces for the prevention of bacterial/protein attachment is a most commonly used approach for the production of antimicrobial surfaces (Banerjee et al. 2011). Otsuka et al. provided a detailed review of PEG-based block copolymers and their applications (Otsuka et al. 2001). One study using the PEG-graft-poly(acrylic acid) (PEG-g-PAA) copolymer showed that inhibition of non-specific protein and bacterial adhesion on surfaces could be achieved. These authors further modified these copolymers with peptides and fragments of antibodies specific to macrophage cell-surface proteins to investigate the antifouling capability of such surfaces against three different proteins; human serum albumin (HSA), human fibrinogen (FGN), and human immunoglobulin (IgG). Modification studies showed that peptides linked with star PEO and star-like PEO tethers reduced the extent of non-specific protein adsorption by 40–92 % compared to the controls. In addition, these authors used ^{14}C-labeled *S. epidermis* and *P. aeruginosa* pathogens to quantify the degree of adhesion, with some contrasting results being shown. For example, both bacterial species showed lower attachment on both peptide-modified and unmodified surfaces. There was no such effect on adhesion using directly bound or PEO bound peptides on *P. aeruginosa*, while *S. epidermis* adhesion increased a little on surfaces where peptides were directly attached to the co-polymer, compared with peptides bound with PEO tethers (Wagner et al. 2004). Another study demonstrated the compression of solvated PEG brushes on coated surfaces during the initial attachment of *S. aureus*. A ~10 nm range cationic polymer patch was placed at the base of PEG brushes, which was electrostatically attractive towards protein and the negatively charged bacterium *S. aureus*. The key finding was that the mass or chain length of the PEG molecules was the main factor responsible for the repulsion of bacteria, with the prevention of protein adsorption requiring a greater number of cationic patches compared to that required by the bacteria. Protein adsorption was also dependent on the structure of the PEG because small protein molecules could easily penetrate into the gaps present in the chain structure. These results indicated that polymer brush based protein resistant surfaces have nanoscale defects that can induce bacterial fouling even before proteins adsorb (Gon et al. 2012). Campoccia et al. published a detailed review of different polymer coatings that resist bacterial adhesion (Campoccia et al. 2013).

5.4.4 Bactericidal Release Systems from Implants Surfaces

The starting point of pathogenesis is the initial attachment of bacteria on the surface of implants. A great deal of research has focussed on the development of biocompatible polymer coatings on implants that can prevent this process; this is made possible due to the physiochemical modification or incorporation of antibacterial agents into the biocompatible polymers. Combination strategies show a great deal of promise, because they can reduce the initial bacterial attachment via the active release of antibacterial compounds from implant surfaces. These types of active surfaces are able to release a strong initial concentration of antibacterial agents within the crucial initial short term period post-implantation to inhibit the initial attachment of bacteria (Hetrick and Schoenfisch 2006). Hendricks et al. used poly(etherurethane) (PEU) as a base material mixed with powdered PEO and the antibiotic ciprofloxacin™ to form an antimicrobial surface. This surface was further modified by the plasma polymerisation of poly(butyl methacrylate) (pBMA) onto the surface to serve as a porous barrier layer for controlling the release of the antibacterial compound. Twenty four hour incubation with a *P. aeruginosa* suspension showed that a significantly lower amount of biofilm development occurred compared to the control surface (Hendricks et al. 2000). Also, Gutierrez et al. used low density polypropylene meshes coated with a biocompatible and resorbable polymer (non-cross-linked copolymer of 2-hydroxyethyl methacrylate and 2-acrylamido-2-methylpropanesulfonic acid) (HEMA-AMPS) for the controlled release of antibiotics. The coating was resorbed in the media due to slow dissolution and able to retain its activity for a minimum of 30 days post-implantation in rabbits. In addition, significantly lower concentrations of the antibiotic (vancomycin) were detected in the blood stream, which proved the drug was mainly released locally (Fernandez-Gutierrez et al. 2013). The main advantage of resorbable polymers it that during resorption, the antibacterial compounds are released and the activity of the coatings can be tuned by the degradation rates of the polymer and thickness of the coatings. Shukla et al. (2010) took advantage of LbL assembly of polymer thin films to incorporate AMPs (Izadpanah and Gallo 2005; Hancock and Diamond 2000), with the main focus being the natural form of AMP, i.e. ponericin G1. The results showed promising inhibition of *S. aureus* attachment with control of the drug loading, release kinetics and also the biocompatibility of polymers when used with wound healing cells such as NIH 3T3 embryonic murine fibroblasts and human umbilical vein endothelial cells (HUVECs). In another study, these authors incorporated gentamicin during the LbL assembly of hydrolytically degradable polyelectrolyte polymers such as poly (β-amino ester) and PAA, which gets released in contact with the aqueous medium through both film erosion and diffusion. The resultant bactericidal affects showed an average of 5 orders of magnitude reduction in *S. aureus* adhesion. No toxicity was shown towards the MC3T3-E1 murine preosteoblasts cells compared to previous studies (Isefuku et al. 2003), which showed that a high concentration of gentamicin affects the extent of osteoblast cell proliferation. From this comparison, it may be concluded that this LbL coating is releasing low concentrations of gentamicin. A subsequent *in vivo* study of polyelectrolyte coated titanium in New Zealand white rabbits resulted in a significant decrease in the viable bacteria count compared to that of the non-coated implants (Moskowitz et al. 2010). LbL polymer coatings have been shown to deliver multiple drugs, including antibiotic and non-steroidal anti-inflammatory drugs, for controlling infection and inflammation (Shukla et al. 2011). Gentamicin is an effective antibiotic against *P. aeruginosa* but its uses are tempered due to its short half-life, lower bioavailability and cytotoxicity (Tange et al. 1995; Abdelghany et al. 2012). To overcome this problem, Abdelghany et al. developed a control release system for gentamicin using nanoparticles made up of poly(lactide-co-glycolide) (PLGA). Their work resulted in the development of PLGA formulations of gentamicin that exhibited longer lasting antimicrobial effects against *P. aeruginosa* compared to that obtained using free gentamicin *in vivo* (Abdelghany et al. 2012).

5.4.5 Self-Cleaning Surfaces

In the last few years, bio-inspired superhydrophobic surfaces such as those found on lotus leaves, cicada wings, shark and fish skin, and *Nelumbo nucifera* have drawn attention as potential methods by which biofilm formation can be eradicated. Mimicking of the surface structure of these natural organisms represent a novel method for the design of antifouling surfaces. Khalil-bad et al. produced superhydrophobic antibacterial cotton via the incorporation of Ag nanoparticles into the cotton surface, followed by dip-coating in octyltriethoxysilane (OTES) solution (Shateri Khalil-Abad and Yazdanshenas 2010). Their results showed that the cotton increased its water contact angle to 151°, displaying antibacterial activity against both Gram negative (*E. coli*) and Gram positive bacteria (*S. aureus*). In another study, superhydrophobic surfaces could be converted to the Cassie-Baxter state (a surface state in which air is trapped in the interstices on the surface, forming a hydrophobic (solid/air) interface, which resulted in a higher contact angle compared to that of a flat surface of the same material), which resulted in reduced contact taking place between bacteria and the surfaces. The surface was very rough and synthesized from a silicone elastomer with the help of an aerosol assisted chemical vapour deposition (AACVD) process. The behaviour of this surface resembled that of the lotus leaf, with water contact angles of approximately 165° being obtained. *E. coli* and methicillin-resistant *S. aureus* attachment were tested and it was shown that 79 % less attachment took place on these surfaces compared to the control surface of planar glass (Crick et al. 2011). Loo et al. demonstrated a delay in *P. aeruginosa* attachment to incubation tubes and medical plastic comprised of superhydrophobic poly(vinyl chloride) (PVC) surfaces (Loo et al. 2012). These authors modified the PVC surface with a combination of the solvent tetrahydrofuran and non-solvents such as ethanol and methanol. Ethanol treated surfaces showed more hydrophobicity compared to that of the methanol treated surfaces. As the concentration of ethanol was increased from 15 to 35 %, the water contact angle increased from 73° to 150° due to changes in the surface architecture. The colonization study using *P. aeruginosa* showed that 18–24 h delays in attachment could be achieved compared to that of the control, unmodified PVC. The concept of superhydrophobic surfaces being an effective tool in delaying the initial bacterial attachment to surfaces is thus plausible. It has been shown that superhydrophobic surfaces effectively reduce bacterial attachment on surfaces; however a detailed investigation of the characteristics responsible for inhibition of bacterial attachment is yet to be performed. To understand this phenomenon, Yoon et al. compared the amount of bacteria adhering to superhydrophobic and superhydrophilic surfaces under different fluid flow conditions (Yoon et al. 2014). They separately synthesized both superhydrophobic and superhydrophilic surfaces by annealing stainless steel plates with carbon nanotubes-poly (tetrafluoroethylene) (CNT-PTFE) and titanium dioxide (TiO_2). *E. coli* K-12 (3×10^8 cells/mL) suspended in PBS buffer were pumped through the chamber at flow rates of both zero (0) and at 200 mL/min, hence studying the effect of surface shear on the attachment process. Their result highlighted that microbial adhesion rates on substrata surfaces could be influenced by the shear or flow conditions, and surface morphologies. The super-hydrophobic CNT-PTFE (water contact angle of 154.6°) coating showed much lower levels of adherent bacteria to the highly sheared surface because of its lotus leaf effect. In contrast, superhydrophilic surfaces showed comparable an anti-adhesive property in a stagnant environment due the presence of a chemically bound hydration layer. These results suggested that surfaces that exhibit anti-adhesion properties in both stagnant and continue flow conditions are needed.

5.4.6 Novel Strategies to Control Biofilm Formation

There are a number of emerging strategies to combat microbial derived infection other than those involving surface modification. Hazan et al. used low energy surface acoustic waves (SAW) to prevent medical device infection, with the example being that urinary catheters could be rendered sterile for up to 9 days compared to the control samples that were not subjected to the acoustic waves (Hazan et al. 2006). They speculated that the chaotic microstreaming generated in fluids by the vibrations arising from the application of the acoustic waves disrupted the quorum sensing mechanisms of the bacteria. Continuous low energy SAW was transmitted directly to the extracorporeal portion of implanted device. Disruption of the SAW was found to promote renewed adhesion of bacteria. To explain the biofilm prevention properties of the application of SAWs, these authors proposed the following explanation: Attraction or repulsion of bacteria in the 10 nm range near the surface is an outcome of the van der Waals and hydrophobic attraction forces was counteracted by electrostatic repulsion. Thus, the application of SAWs was effective in delaying the onset of biofilm formation. In can be envisaged that controlling the interaction forces at surfaces by the application of an external stimuli, in combination with other surface modification strategies, could be used to delay the onset of biofilm formation. From a clinical perspective, slowing the implant derived infection has significant benefits particularly when short-term implants such as catheters are being used.

5.5 Conclusions and Future Perspectives

The control of infection from medical implants remains a major global healthcare problem despite tremendous efforts to solve the problem through the surface modification of implant materials. One challenge in biomaterials science is being able to prevent biofilm formation as bacteria become more resistant to antibiotics. In this chapter, we have attempted to discuss the current state-of-the-art in terms of chemical surface modification approaches for the generation of antibacterial or low adhesion surfaces, but clearly many challenges remain.

Antimicrobial polymers, such cationic polymers, have shown effectiveness against bacterial infection but their long term use introduces toxicity as a concern. PEG/PEO based coatings have shown impressive antifouling behaviour but their long effectiveness remains unknown. Quaternary ammonium/phosphonium-based polymers have shown improved performance in antibacterial behaviour over PEG based polymers, especially *in vivo*, but clearly consolidation of performance is required with support from advances in surface analysis techniques in order to fully understand the effectiveness of the range of different surface modification approaches. Plasma-based surface modification strategies continue to show potential for the generation of antibacterial materials, with the prospect of direct modification using antibacterial agents, or being using as coatings to control their release. Self-cleaning, superhydrophobic surfaces are an additional approach being used for repelling bacteria but the long term performance requires long-term study. Any approach employed to delay biofilm formation and infection will have substantial clinical benefits. A multi-strategy surface modification approach is most likely the way forward for the generation of optimal antimicrobial surfaces in areas where biofilms are a problem, where antifouling/antibacterial surfaces could be a possible solution.

Acknowledgments The Scientific Industrial Endowment Fund (SIEF) is acknowledged for providing a John Stocker Postdoctoral Research Fellowship for PYW. The Australian Research Council is acknowledged for funding a PhD scholarship for HP through an ARC Discovery Grant. This work was supported in part at both the Biointerface Engineering Hub at Swinburne and the Melbourne Centre for Nanofabrication as part of the Victorian Node of the Australian National Fabrication Facility, a company established under the National Collaborative Research Infrastructure Strategy to provide nano and microfabrication facilities for Australia's researchers.

References

Abdelghany SM, Quinn DJ, Ingram RJ, Gilmore BF, Donnelly RF, Taggart CC, Scott CJ (2012) Gentamicin-loaded nanoparticles show improved antimicrobial effects towards Pseudomonas aeruginosa infection. Int J Nanomedicine 7:4053

Al-Badri ZM, Som A, Lyon S, Nelson CF, Nüsslein K, Tew GN (2008) Investigating the effect of increasing charge density on the hemolytic activity of synthetic antimicrobial polymers. Biomacromolecules 9(10):2805–2810

Al-Bataineh SA, Jasieniak M, Britcher LG, Griesser HJ (2008) TOF-SIMS and principal component analysis characterization of the multilayer surface grafting of small molecules: antibacterial furanones. Anal Chem 80(2):430–436

Allegranzi B, Nejad SB, Combescure C, Graafmans W, Attar H, Donaldson L, Pittet D (2011) Burden of endemic health-care-associated infection in developing countries: systematic review and meta-analysis. Lancet 377(9761):228–241

Applerot G, Abu-Mukh R, Irzh A, Charmet J, Keppner H, Laux E, Guibert G, Gedanken A (2010) Decorating parylene-coated glass with ZnO nanoparticles for antibacterial applications: a comparative study of sonochemical, microwave, and microwave-plasma coating routes. ACS Appl Mater Interfaces 2(4):1052–1059

Augustin M, Ali-Vehmas T, Atroshi F (2004) Assessment of enzymatic cleaning agents and disinfectants against bacterial biofilms. J Pharm Pharm Sci 7:55–64

Banerjee I, Pangule RC, Kane RS (2011) Antifouling coatings: recent developments in the design of surfaces that prevent fouling by proteins, bacteria, and marine organisms. Adv Mater 23(6):690–718

Barken KB, Pamp SJ, Yang L, Gjermansen M, Bertrand JJ, Klausen M, Givskov M, Whitchurch CB, Engel JN, Tolker-Nielsen T (2008) Roles of type IV pili, flagellum-mediated motility and extracellular DNA in the formation of mature multicellular structures in Pseudomonas aeruginosa biofilms. Environ Microbiol 10(9):2331–2343

Bazaka K, Jacob MV, Truong VK, Wang F, Pushpamali WAA, Wang JY, Ellis AV, Berndt CC, Crawford RJ, Ivanova EP (2010) Plasma-enhanced synthesis of bioactive polymeric coatings from monoterpene alcohols: a combined experimental and theoretical study. Biomacromolecules 11(8):2016–2026

Bazaka K, Jacob MV, Crawford RJ, Ivanova EP (2011) Plasma-assisted surface modification of organic biopolymers to prevent bacterial attachment. Acta Biomater 7(5):2015–2028

Bazaka K, Jacob MV, Crawford RJ, Ivanova EP (2012) Efficient surface modification of biomaterial to prevent biofilm formation and the attachment of microorganisms. Appl Microbiol Biotechnol 95(2):299–311

Bernbom N, Jørgensen RL, Ng Y, Meyer R, Kingshott P, Vejborg RM, Klemm P, Besenbacher F, Gram L (2006) Bacterial adhesion to stainless steel is reduced by aqueous fish extract coatings. Biofilms 3(01):25–36

Bernbom N, Ng YY, Jørgensen RL, Arpanaei A, Meyer RL, Kingshott P, Vejborg RM, Klemm P, Gram L (2009) Adhesion of food-borne bacteria to stainless steel is reduced by food conditioning films. J Appl Microbiol 106(4):1268–1279

Berne C, Kysela DT, Brun YV (2010) A bacterial extracellular DNA inhibits settling of motile progeny cells within a biofilm. Mol Microbiol 77(4):815–829

Bilek MM, McKenzie DR (2010) Plasma modified surfaces for covalent immobilization of functional biomolecules in the absence of chemical linkers: towards better biosensors and a new generation of medical implants. Biophys Rev 2(2):55–65

Campoccia D, Montanaro L, Arciola CR (2013) A review of the biomaterials technologies for infection-resistant surfaces. Biomaterials 34(34):8533–8554

Carmen JC, Roeder BL, Nelson JL, Robison Ogilvie RL, Robison RA, Schaalje GB, Pitt WG (2005) Treatment of biofilm infections on implants with low-frequency ultrasound and antibiotics. Am J Infect Control 33(2):78–82

Castellsagué X, Díaz M, Vaccarella S, de Sanjosé S, Muñoz N, Herrero R, Franceschi S, Meijer CJ, Bosch FX (2011) Intrauterine device use, cervical infection with human papillomavirus, and risk of cervical cancer: a pooled analysis of 26 epidemiological studies. Lancet Oncol 12(11):1023–1031

Chapman RG, Ostuni E, Liang MN, Meluleni G, Kim E, Yan L, Pier G, Warren HS, Whitesides GM (2001) Polymeric thin films that resist the adsorption of proteins and the adhesion of bacteria. Langmuir 17(4):1225–1233

Chen C, Cooper S (2000) Polycationic antimicrobial dendrimers: a comparison of alkyl pyridinium, quaternary ammonium, quaternary phosphonium and tertiary sulfonium salts. In: APS Meeting abstracts, p 13008

Chen Q, Zhou M, Fu Y, Weng J, Zhang Y, Yue L, Xie F, Huo C (2008) Magnetron sputtering synthesis silver and organic PEO nanocomposite. Surf Coat Technol 202(22–23):5576–5578
Cheng G, Zhang Z, Chen S, Bryers JD, Jiang S (2007) Inhibition of bacterial adhesion and biofilm formation on zwitterionic surfaces. Biomaterials 28(29):4192–4199
Cheng G, Xue H, Zhang Z, Chen S, Jiang S (2008) A switchable biocompatible polymer surface with self-sterilizing and nonfouling capabilities. Angew Chem Int Ed 47(46):8831–8834
Cheng G, Li G, Xue H, Chen S, Bryers JD, Jiang S (2009) Zwitterionic carboxybetaine polymer surfaces and their resistance to long-term biofilm formation. Biomaterials 30(28):5234–5240
Chenoweth C, Saint S (2013) Preventing catheter-associated urinary tract infections in the intensive care unit. Crit Care Clin 29(1):19–32
Chua P-H, Neoh K-G, Kang E-T, Wang W (2008) Surface functionalization of titanium with hyaluronic acid/chitosan polyelectrolyte multilayers and RGD for promoting osteoblast functions and inhibiting bacterial adhesion. Biomaterials 29(10):1412–1421
Conover MS, Mishra M, Deora R (2011) Extracellular DNA is essential for maintaining Bordetella biofilm integrity on abiotic surfaces and in the upper respiratory tract of mice. PLoS ONE 6(2):e16861
Cornell RJ, Donaruma LG (1965) 2-Methacryloxytropones. Intermediates for the synthesis of biologically active polymers. J Med Chem 8(3):388–390
Costa F, Carvalho IF, Montelaro RC, Gomes P, Martins MCL (2011) Covalent immobilization of antimicrobial peptides (AMPs) onto biomaterial surfaces. Acta Biomater 7(4):1431–1440
Cowan MM (1999) Plant products as antimicrobial agents. Clin Microbiol Rev 12(4):564-+
Crick CR, Ismail S, Pratten J, Parkin IP (2011) An investigation into bacterial attachment to an elastomeric superhydrophobic surface prepared via aerosol assisted deposition. Thin Solid Films 519(11):3722–3727
Cucarella C, Solano C, Valle J, Amorena B, Lasa Í, Penadés JR (2001) Bap, a Staphylococcus aureus surface protein involved in biofilm formation. J Bacteriol 183(9):2888–2896
Danese PN, Pratt LA, Dove SL, Kolter R (2000) The outer membrane protein, Antigen 43, mediates cell-to-cell interactions within Escherichia coli biofilms. Mol Microbiol 37(2):424–432
Daniel A, Le Pen C, Archambeau C, Reniers F (2009) Use of a PECVD–PVD process for the deposition of copper containing organosilicon thin films on steel. Appl Surf Sci 256(3, Supplement):S82–S85
Darouiche RO (2004) Treatment of infections associated with surgical implants. N Engl J Med 350(14):1422–1429
Deng L, Mrksich M, Whitesides GM (1996) Self-assembled monolayers of alkanethiolates presenting tri (propylene sulfoxide) groups resist the adsorption of protein. J Am Chem Soc 118(21):5136–5137
Di Poto A, Sbarra MS, Provenza G, Visai L, Speziale P (2009) The effect of photodynamic treatment combined with antibiotic action or host defence mechanisms on Staphylococcus aureus biofilms. Biomaterials 30(18):3158–3166
Ding X, Yang C, Lim TP, Hsu LY, Engler AC, Hedrick JL, Yang Y-Y (2012) Antibacterial and antifouling catheter coatings using surface grafted PEG-b-cationic polycarbonate diblock copolymers. Biomaterials 33(28):6593–6603
Dizman B, Elasri MO, Mathias LJ (2004) Synthesis, characterization, and antibacterial activities of novel methacrylate polymers containing norfloxacin. Biomacromolecules 6(1):514–520
Donlan RM, Costerton JW (2002) Biofilms: survival mechanisms of clinically relevant microorganisms. Clin Microbiol Rev 15(2):167–193
Dunne WM (2002) Bacterial adhesion: seen any good biofilms lately? Clin Microbiol Rev 15(2):155–166
Dutta D, Cole N, Willcox M (2012) Factors influencing bacterial adhesion to contact lenses. Mol Vis 18:14
Fernandez-Gutierrez M, Olivares E, Pascual G, Bellon JM, Román JS (2013) Low-density polypropylene meshes coated with resorbable and biocompatible hydrophilic polymers as controlled release agents of antibiotics. Acta Biomater 9(4):6006–6018

Fey PD (2010) Modality of bacterial growth presents unique targets: how do we treat biofilm-mediated infections? Curr Opin Microbiol 13(5):610–615

Frieden T (2013) Antibiotic resistance threats in the United States. Centers for Disease Control and Prevention, Atlanta

Fuglsang CC, Johansen C, Christgau S, Adler-Nissen J (1995) Antimicrobial enzymes: applications and future potential in the food industry. Trends Food Sci Technol 6(12):390–396

Fundeanu I, van der Mei HC, Schouten AJ, Busscher HJ (2008) Polyacrylamide brush coatings preventing microbial adhesion to silicone rubber. Colloids Surf B: Biointerfaces 64(2):297–301

Garcia-Fernandez MJ, Martinez-Calvo L, Ruiz JC, Wertheimer MR, Concheiro A, Alvarez-Lorenzo C (2012) Loading and release of drugs from oxygen-rich plasma polymer coatings. Plasma Process Polym 9(5):540–549

Gloag ES, Turnbull L, Huang A, Vallotton P, Wang H, Nolan LM, Mililli L, Hunt C, Lu J, Osvath SR (2013) Self-organization of bacterial biofilms is facilitated by extracellular DNA. Proc Natl Acad Sci 110(28):11541–11546

Gon S, Kumar K-N, Nüsslein K, Santore MM (2012) How bacteria adhere to brushy peg surfaces: clinging to flaws and compressing the brush. Macromolecules 45(20):8373–8381

Goy RC, Dd B, Assis OB (2009) A review of the antimicrobial activity of chitosan. Polímeros 19(3):241–247

Griffith BP, Kormos RL, Hardesty RL, Armitage JM, Dummer JS (1988) The artificial heart: infection-related morbidity and its effect on transplantation. Ann Thorac Surg 45(4):409–414

Grigoras AG, Racovita S, Vasiliu S, Nistor MT, Dunca S, Barboiu V, Grigoras VC (2012) Dilute solution properties of some polycarboxybetaines with antibacterial activity. J Polym Res 19(11):8

Gudipati CS, Finlay JA, Callow JA, Callow ME, Wooley KL (2005) The antifouling and fouling-release performance of hyperbranched fluoropolymer (HBFP)-poly (ethylene glycol)(PEG) composite coatings evaluated by adsorption of biomacromolecules and the green fouling alga ulva. Langmuir 21(7):3044–3053

Guo L, Yuan W, Lu Z, Li CM (2013) Polymer/nanosilver composite coatings for antibacterial applications. Colloids Surf A Physicochem Eng Asp 439:69–83

Hancock REW, Diamond G (2000) The role of cationic antimicrobial peptides in innate host defences. Trends Microbiol 8(9):402–410

Hancock REW, Sahl H-G (2006) Antimicrobial and host-defense peptides as new anti-infective therapeutic strategies. Nat Biotech 24(12):1551–1557

Harmsen M, Lappann M, Knøchel S, Molin S (2010) Role of extracellular DNA during biofilm formation by Listeria monocytogenes. Appl Environ Microbiol 76(7):2271–2279

Harris LG, Tosatti S, Wieland M, Textor M, Richards RG (2004) Staphylococcus aureus adhesion to titanium oxide surfaces coated with non-functionalized and peptide-functionalized poly(l-lysine)-grafted-poly(ethylene glycol) copolymers. Biomaterials 25(18):4135–4148

Hasan J, Crawford RJ, Ivanova EP (2013) Antibacterial surfaces: the quest for a new generation of biomaterials. Trends Biotechnol 31(5):295–304

Hazan Z, Zumeris J, Jacob H, Raskin H, Kratysh G, Vishnia M, Dror N, Barliya T, Mandel M, Lavie G (2006) Effective prevention of microbial biofilm formation on medical devices by low-energy surface acoustic waves. Antimicrob Agents Chemother 50(12):4144–4152

He D, Noh SK, Lyoo WS (2011) In situ-generated Ru (III)-mediated ATRP from the polymeric Ru (III) complex in the absence of activator generation agents. J Polym Sci A Polym Chem 49(21):4594–4602

Heilmann C, Hussain M, Peters G, Götz F (1997) Evidence for autolysin-mediated primary attachment of Staphylococcus epidermidis to a polystyrene surface. Mol Microbiol 24(5):1013–1024

Heimberger T, Duma R (1989) Infections of prosthetic heart valves and cardiac pacemakers. Infect Dis Clin N Am 3(2):221–245

Hendricks SK, Kwok C, Shen M, Horbett TA, Ratner BD, Bryers JD (2000) Plasma-deposited membranes for controlled release of antibiotic to prevent bacterial adhesion and biofilm formation. J Biomed Mater Res 50(2):160–170

Hetrick EM, Schoenfisch MH (2006) Reducing implant-related infections: active release strategies. Chem Soc Rev 35(9):780–789

Hou S, Burton EA, Simon KA, Blodgett D, Luk Y-Y, Ren D (2007) Inhibition of Escherichia coli biofilm formation by self-assembled monolayers of functional alkanethiols on gold. Appl Environ Microbiol 73(13):4300–4307

Iarikov DD, Kargar M, Sahari A, Russel L, Gause KT, Behkam B, Ducker WA (2013) Antimicrobial surfaces using covalently bound polyallylamine. Biomacromolecules 15(1):169–176

Isefuku S, Joyner CJ, Simpson AHR (2003) Gentamicin may have an adverse effect on osteogenesis. J Orthop Trauma 17(3):212–216

Izadpanah A, Gallo RL (2005) Antimicrobial peptides. J Am Acad Dermatol 52(3):381–390

Izano EA, Amarante MA, Kher WB, Kaplan JB (2008) Differential roles of poly-N-acetylglucosamine surface polysaccharide and extracellular DNA in Staphylococcus aureus and Staphylococcus epidermidis biofilms. Appl Environ Microbiol 74(2):470–476

Jacobs T, Morent R, De Geyter N, Dubruel P, Leys C (2012) Plasma surface modification of biomedical polymers: influence on cell-material interaction. Plasma Chem Plasma Process 32(5):1039–1073

Jahani S, Shakiba A, Azami M (2014) Functional properties, antibacterial and antioxidant activities of Zataria multiflora encapsulated in geltin nanofilms. J Microbiol Biotechnol Food Sci 4(2):88–92

Jiang B, Barnett JB, Li B (2009) Advances in polyelectrolyte multilayer nanofilms as tunable drug delivery systems. Nanotechnol Sci Appl 2:21

Jones N, Ray B, Ranjit KT, Manna AC (2008) Antibacterial activity of ZnO nanoparticle suspensions on a broad spectrum of microorganisms. FEMS Microbiol Lett 279(1):71–76

Kalemba D, Kunicka A (2003) Antibacterial and antifungal properties of essential oils. Curr Med Chem 10(10):813–829

Kanazawa A, Ikeda T, Endo T (1993a) Polymeric phosphonium salts as a novel class of cationic biocides. IV. Synthesis and antibacterial activity of polymers with phosphonium salts in the main chain. J Polym Sci A Polym Chem 31(12):3031–3038

Kanazawa A, Ikeda T, Endo T (1993b) Novel polycationic biocides: synthesis and antibacterial activity of polymeric phosphonium salts. J Polym Sci A Polym Chem 31(2):335–343

Kawabata N, Nishiguchi M (1988) Antibacterial activity of soluble pyridinium-type polymers. Appl Environ Microbiol 54(10):2532–2535

Kenawy E-R, Worley S, Broughton R (2007) The chemistry and applications of antimicrobial polymers: a state-of-the-art review. Biomacromolecules 8(5):1359–1384

Kim JS, Kuk E, Yu KN, Kim J-H, Park SJ, Lee HJ, Kim SH, Park YK, Park YH, Hwang C-Y, Kim Y-K, Lee Y-S, Jeong DH, Cho M-H (2007) Antimicrobial effects of silver nanoparticles. Nanomed Nanotechnol Biol Med 3(1):95–101

Kingshott P, Griesser HJ (1999) Surfaces that resist bioadhesion. Curr Opinion Solid State Mater Sci 4(4):403–412

Kingshott P, Thissen H, Griesser HJ (2002) Effects of cloud-point grafting, chain length, and density of PEG layers on competitive adsorption of ocular proteins. Biomaterials 23(9):2043–2056

Kingshott P, Wei J, Bagge-Ravn D, Gadegaard N, Gram L (2003) Covalent attachment of poly(ethylene glycol) to surfaces, critical for reducing bacterial adhesion. Langmuir 19(17): 6912–6921

Klausen M, Aaes-Jørgensen A, Molin S, Tolker-Nielsen T (2003) Involvement of bacterial migration in the development of complex multicellular structures in Pseudomonas aeruginosa biofilms. Mol Microbiol 50(1):61–68

Klevens RM, Edwards JR, Richards CL, Horan TC, Gaynes RP, Pollock DA, Cardo DM (2007) Estimating health care-associated infections and deaths in US hospitals, 2002. Public Health Rep 122(2):160

Kwok CS, Horbett TA, Ratner BD (1999) Design of infection-resistant antibiotic-releasing polymers: II. Controlled release of antibiotics through a plasma-deposited thin film barrier. J Control Release 62(3):301–311

Kwon D-H, Lu C-D (2007) Polyamine effects on antibiotic susceptibility in bacteria. Antimicrob Agents Chemother 51(6):2070–2077

Lee SB, Koepsel RR, Morley SW, Matyjaszewski K, Sun Y, Russell AJ (2004) Permanent, non-leaching antibacterial surfaces. 1. Synthesis by atom transfer radical polymerization. Biomacromolecules 5(3):877–882

Leroy C, Delbarre C, Ghillebaert F, Compere C, Combes D (2008) Effects of commercial enzymes on the adhesion of a marine biofilm-forming bacterium. Biofouling 24(1):11–22

Li G, Cheng G, Xue H, Chen S, Zhang F, Jiang S (2008a) Ultra low fouling zwitterionic polymers with a biomimetic adhesive group. Biomaterials 29(35):4592–4597

Li G, Xue H, Cheng G, Chen S, Zhang F, Jiang S (2008b) Ultralow fouling zwitterionic polymers grafted from surfaces covered with an initiator via an adhesive mussel mimetic linkage. J Phys Chem B 112(48):15269–15274

Li P, Poon YF, Li W, Zhu H-Y, Yeap SH, Cao Y, Qi X, Zhou C, Lamrani M, Beuerman RW, Kang E-T, Mu Y, Li CM, Chang MW, Jan Leong SS, Chan-Park MB (2011) A polycationic antimicrobial and biocompatible hydrogel with microbe membrane suctioning ability. Nat Mater 10(2):149–156

Lin J, Tiller JC, Lee SB, Lewis K, Klibanov AM (2002) Insights into bactericidal action of surface-attached poly (vinyl-N-hexylpyridinium) chains. Biotechnol Lett 24(10):801–805

Lin J, Qiu S, Lewis K, Klibanov AM (2003) Mechanism of bactericidal and fungicidal activities of textiles covalently modified with alkylated polyethylenimine. Biotechnol Bioeng 83(2):168–172

Loo C-Y, Young PM, Lee W-H, Cavaliere R, Whitchurch CB, Rohanizadeh R (2012) Superhydrophobic, nanotextured polyvinyl chloride films for delaying Pseudomonas aeruginosa attachment to intubation tubes and medical plastics. Acta Biomater 8(5):1881–1890

Lyte M, Freestone PP, Neal CP, Olson BA, Haigh RD, Bayston R, Williams PH (2003) Stimulation of Staphylococcus epidermidis growth and biofilm formation by catecholamine inotropes. Lancet 361(9352):130–135

Ma Y, Chen M, Jones JE, Ritts AC, Yu Q, Sun H (2012) Inhibition of Staphylococcus epidermidis biofilm by trimethylsilane plasma coating. Antimicrob Agents Chemother 56(11):5923–5937

Majik MS, Parvatkar TP (2014) Next generation biofilm inhibitors for Pseudomonas aeruginosa: synthesis and rational design approaches. Curr Top Med Chem 14(1):81–109

Manne MB, Shrestha NK, Lytle BW, Nowicki ER, Blackstone E, Gordon SM, Pettersson G, Fraser TG (2012) Outcomes after surgical treatment of native and prosthetic valve infective endocarditis. Ann Thorac Surg 93(2):489–493

Moore WS, Chvapil M, Seiffert G, Keown K (1981) Development of an infection-resistant vascular prosthesis. Arch Surg 116(11):1403–1407

Morra M, Cassineli C (1999) Non-fouling properties of polysaccharide-coated surfaces. J Biomater Sci Polym Ed 10(10):1107–1124

Moskowitz JS, Blaisse MR, Samuel RE, Hsu H-P, Harris MB, Martin SD, Lee JC, Spector M, Hammond PT (2010) The effectiveness of the controlled release of gentamicin from polyelectrolyte multilayers in the treatment of Staphylococcus aureus infection in a rabbit bone model. Biomaterials 31(23):6019–6030

Muñoz-Bonilla A, Fernández-García M (2012) Polymeric materials with antimicrobial activity. Prog Polym Sci 37(2):281–339

Nablo BJ, Rothrock AR, Schoenfisch MH (2005) Nitric oxide-releasing sol–gels as antibacterial coatings for orthopedic implants. Biomaterials 26(8):917–924

Nguyen LT, Haney EF, Vogel HJ (2011) The expanding scope of antimicrobial peptide structures and their modes of action. Trends Biotechnol 29(9):464–472

O'Toole GA, Kolter R (1998) Flagellar and twitching motility are necessary for Pseudomonas aeruginosa biofilm development. Mol Microbiol 30(2):295–304

Oren Z, Shai Y (1998) Mode of action of linear amphipathic α-helical antimicrobial peptides. Pept Sci 47(6):451–463

Ortlepp S, Pedpradap S, Dobretsov S, Proksch P (2008) Antifouling activity of sponge-derived polybrominated diphenyl ethers and synthetic analogues. Biofouling 24(3):201–208

Otsuka H, Nagasaki Y, Kataoka K (2001) Self-assembly of poly (ethylene glycol)-based block copolymers for biomedical applications. Curr Opin Colloid Interface Sci 6(1):3–10

Percival SL, Walker JT, Hunter PR (2000) Microbiological aspects of biofilms and drinking water. CRC Press, New York

Percival S, Malic S, Cruz H, Williams D (2011) Introduction to biofilms. In: Percival S, Knottenbelt D, Cochrane C (eds) Biofilms and veterinary medicine, vol 6. Springer, Berlin/Heidelberg, pp 41–68

Petersen FC, Tao L, Scheie AA (2005) DNA binding-uptake system: a link between cell-to-cell communication and biofilm formation. J Bacteriol 187(13):4392–4400

Pillai VR, Mutter M (1981) Synthetic hydrophilic polymers. Naturwissenschaften 68(11):558–566

Poelstra KA, Barekzi NA, Rediske AM, Felts AG, Slunt JB, Grainger DW (2002) Prophylactic treatment of gram-positive and gram-negative abdominal implant infections using locally delivered polyclonal antibodies. J Biomed Mater Res 60(1):206–215

Poortinga AT, Bos R, Norde W, Busscher HJ (2002) Electric double layer interactions in bacterial adhesion to surfaces. Surf Sci Rep 47(1):1–32

Popat KC, Eltgroth M, LaTempa TJ, Grimes CA, Desai TA (2007) Decreased Staphylococcus epidermis adhesion and increased osteoblast functionality on antibiotic-loaded titania nanotubes. Biomaterials 28(32):4880–4888

Post JC, Stoodley P, Hall–Stoodley L, Ehrlich GD (2004) The role of biofilms in otolaryngologic infections. Curr Opin Otolaryngol Head Neck Surg 12(3):185–190

Rhim J-W, Hong S-I, Park H-M, Ng PKW (2006) Preparation and characterization of Chitosan-based nanocomposite films with antimicrobial activity. J Agric Food Chem 54(16):5814–5822

Richards MJ, Edwards JR, Culver DH, Gaynes RP (2000) Nosocomial infections in combined medical-surgical intensive care units in the United States. Infect Control Hosp Epidemiol 21(8):510–515

Rodrigues, L. R. (2011). Inhibition of bacterial adhesion on medical devices. In: Linke D, Goldman A (eds) Bacterial adhesion, vol. 715. Springer, p 351)

Rosenthal VD (2011) Health-care-associated infections in developing countries. Lancet 377:186–188

Ruparelia JP, Chatterjee AK, Duttagupta SP, Mukherji S (2008) Strain specificity in antimicrobial activity of silver and copper nanoparticles. Acta Biomater 4(3):707–716

Senaratne W, Andruzzi L, Ober CK (2005) Self-assembled monolayers and polymer brushes in biotechnology: current applications and future perspectives. Biomacromolecules 6(5):2427–2448

Shao C-L, Wu H-X, Wang C-Y, Liu Q-A, Xu Y, Wei M-Y, Qian P-Y, Gu Y-C, Zheng C-J, She Z-G, Lin Y-C (2011) Potent antifouling resorcylic acid lactones from the Gorgonian-derived fungus Cochliobolus lunatus. J Nat Prod 74(4):629–633

Shateri Khalil-Abad M, Yazdanshenas ME (2010) Superhydrophobic antibacterial cotton textiles. J Colloid Interface Sci 351(1):293–298

Shukla A, Fleming KE, Chuang HF, Chau TM, Loose CR, Stephanopoulos GN, Hammond PT (2010) Controlling the release of peptide antimicrobial agents from surfaces. Biomaterials 31(8):2348–2357

Shukla A, Fuller RC, Hammond PT (2011) Design of multi-drug release coatings targeting infection and inflammation. J Control Release 155(2):159–166
Siedenbiedel F, Tiller JC (2012) Antimicrobial polymers in solution and on surfaces: overview and functional principles. Polymers 4(1):46–71
Sin M-C, Sun Y-M, Chang Y (2014) Zwitterionic-based stainless steel with well-defined polysulfobetaine brushes for general bioadhesive control. ACS Appl Mater Interfaces 6(2):861–873
Smith AW (2005) Biofilms and antibiotic therapy: is there a role for combating bacterial resistance by the use of novel drug delivery systems? Adv Drug Deliv Rev 57(10):1539–1550
Spoering AL, Gilmore MS (2006) Quorum sensing and DNA release in bacterial biofilms. Curr Opin Microbiol 9(2):133–137
Stewart PS, William Costerton J (2001) Antibiotic resistance of bacteria in biofilms. Lancet 358(9276):135–138
Stoodley P, Sauer K, Davies D, Costerton JW (2002) Biofilms as complex differentiated communities. Ann Rev Microbiol 56(1):187–209
Sutherland IW (2001) Biofilm exopolysaccharides: a strong and sticky framework. Microbiology 147(1):3–9
Szleifer I (1997) Protein adsorption on tethered polymer layers: effect of polymer chain architecture and composition. Physica A: Stat Mech Appl 244(1):370–388
Tange R, Dreschler W, Prins J, Buller H, Kuijper E, Speelman P (1995) Ototoxicity and nephrotoxicity of gentamicin vs netilmicin in patients with serious infections. A randomized clinical trial. Clin Otolaryngol Allied Sci 20(2):118–123
Tetz VV, Tetz GV (2010) Effect of extracellular DNA destruction by DNase I on characteristics of forming biofilms. DNA Cell Biol 29(8):399–405
Thallinger B, Prasetyo EN, Nyanhongo GS, Guebitz GM (2013) Antimicrobial enzymes: an emerging strategy to fight microbes and microbial biofilms. Biotechnol J 8(1):97–109
Tiller JC, Liao C-J, Lewis K, Klibanov AM (2001) Designing surfaces that kill bacteria on contact. Proc Natl Acad Sci 98(11):5981–5985
Timofeeva L, Kleshcheva N (2011) Antimicrobial polymers: mechanism of action, factors of activity, and applications. Appl Microbiol Biotechnol 89(3):475–492
Toté K, Berghe DV, Deschacht M, de Wit K, Maes L, Cos P (2009) Inhibitory efficacy of various antibiotics on matrix and viable mass of Staphylococcus aureus and Pseudomonas aeruginosa biofilms. Int J Antimicrob Agents 33(6):525–531
Trentin DS, Bonatto F, Zimmer KR, Ribeiro VB, Antunes ALS, Barth AL, Soares GV, Krug C, Baumvol IJR, Macedo AJ (2014) N2/H2 plasma surface modifications of polystyrene inhibit the adhesion of multidrug resistant bacteria. Surf Coat Technol 245:84–91
Triandafillu K, Balazs D, Aronsson B-O, Descouts P, Tu Quoc P, van Delden C, Mathieu H, Harms H (2003) Adhesion of Pseudomonas aeruginosa strains to untreated and oxygen-plasma treated poly (vinyl chloride)(PVC) from endotracheal intubation devices. Biomaterials 24(8):1507–1518
Vacheethasanee K, Marchant RE (2000) Surfactant polymers designed to suppress bacterial (Staphylococcus epidermidis) adhesion on biomaterials. J Biomed Mater Res 50(3):302–312
Vallet I, Diggle SP, Stacey RE, Cámara M, Ventre I, Lory S, Lazdunski A, Williams P, Filloux A (2004) Biofilm formation in Pseudomonas aeruginosa: fimbrial cup gene clusters are controlled by the transcriptional regulator MvaT. J Bacteriol 186(9):2880–2890
van Schaik EJ, Giltner CL, Audette GF, Keizer DW, Bautista DL, Slupsky CM, Sykes BD, Irvin RT (2005) DNA binding: a novel function of Pseudomonas aeruginosa type IV pili. J Bacteriol 187(4):1455–1464
Vasilev K, Sah V, Anselme K, Ndi C, Mateescu M, Dollmann B, Martinek P, Ys H, Ploux L, Griesser HJ (2009) Tunable antibacterial coatings that support mammalian cell growth. Nano Lett 10(1):202–207
Vasilev K, Griesser SS, Griesser HJ (2011) Antibacterial surfaces and coatings produced by plasma techniques. Plasma Process Polym 8(11):1010–1023

Vilain S, Pretorius JM, Theron J, Brözel VS (2009) DNA as an adhesin: bacillus cereus requires extracellular DNA to form biofilms. Appl Environ Microbiol 75(9):2861–2868

Waar K, van der Mei HC, Harmsen HJ, Degener JE, Busscher HJ (2002) Enterococcus faecalis surface proteins determine its adhesion mechanism to bile drain materials. Microbiology 148(6):1863–1870

Wagner VE, Koberstein JT, Bryers JD (2004) Protein and bacterial fouling characteristics of peptide and antibody decorated surfaces of PEG-poly (acrylic acid) co-polymers. Biomaterials 25(12):2247–2263

Waschinski CJ, Herdes V, Schueler F, Tiller JC (2005) Influence of satellite groups on telechelic antimicrobial functions of polyoxazolines. Macromol Biosci 5(2):149–156

Whitchurch CB, Tolker-Nielsen T, Ragas PC, Mattick JS (2002) Extracellular DNA required for bacterial biofilm formation. Science 295(5559):1487–1487

Wiencek KM, Fletcher M (1995) Bacterial adhesion to hydroxyl-and methyl-terminated alkanethiol self-assembled monolayers. J Bacteriol 177(8):1959–1966

Yamamoto K, Arai H, Ishii M, Igarashi Y (2012) Involvement of flagella-driven motility and pili in Pseudomonas aeruginosa colonization at the air-liquid interface. Microbes Environ 27(3):320–323

Yoon SH, Rungraeng N, Song W, Jun S (2014) Superhydrophobic and superhydrophilic nanocomposite coatings for preventing Escherichia coli K-12 adhesion on food contact surface. J Food Eng 131:135–141

Zasloff M (2002) Antimicrobial peptides of multicellular organisms. Nature 415(6870):389–395

Zegans ME, Becker HI, Budzik J, O'Toole G (2002) The role of bacterial biofilms in ocular infections. DNA Cell Biol 21(5–6):415–420

Zhao L, Mitomo H, Zhai M, Yoshii F, Nagasawa N, Kume T (2003) Synthesis of antibacterial PVA/CM-chitosan blend hydrogels with electron beam irradiation. Carbohydr Polym 53(4):439–446

Chapter 6
Designing Antibacterial Surfaces for Biomedical Implants

Vy T.H. Pham, Chris M. Bhadra, Vi Khanh Truong, Russell J. Crawford, and Elena P. Ivanova

Abstract The infection of biomaterials, particularly medical implants, represents a significant challenge during surgical implantation processes and the subsequent recovery period for the recipient of the implant.

Infections arising from such surgical procedures not only adversely affect the well-being of the patient; they also place a significant burden on the healthcare systems of many countries around the world. A great deal of effort has been made in attempts to minimise or prevent pathogenic bacteria from contaminating these biomaterials. These efforts have included the development of techniques for rendering the surfaces anti-fouling through chemical modification or functionalization of the surface. Recent focus, however, has been placed on the production of antibacterial surfaces. Developments in the area of nanofabrication have allowed the chemical and physical characteristics of the surface of implant materials to be modified such that the molecular to micro-scale topological features can now be accurately controlled.

This chapter will provide an overview of the current approaches and techniques being used or are being developed in the design of antibacterial metallic implant surfaces. Such surfaces can be subjected to a number of chemical and physical modification techniques to achieve this aim, with the resulting surfaces being found to not only exhibit antibacterial behaviour, but also biocompatibility.

Keywords Implant surfaces • Antibacterial • Bactericidal • Titanium • Biocide-releasing surfaces • Nanoparticles • Surface topography

---

V.T.H. Pham • C.M. Bhadra • V.K. Truong • R.J. Crawford • E.P. Ivanova (✉)
School of Science, Faculty of Science, Engineering and Technology,
Swinburne University of Technology, P.O. Box 218, Hawthorn 3122, Victoria, Australia
e-mail: eivanova@swin.edu.au

E.P. Ivanova, R.J. Crawford (eds.), *Antibacterial Surfaces*,
DOI 10.1007/978-3-319-18594-1_6

6.1 Introduction

Bacterial infection of biomaterials has long been a significant problem, both during implant surgery and after the implantation process (Lavernia et al. 2006; Bozic and Ries 2005; Montanaro et al. 2011; Arciola et al. 2012; Campoccia et al. 2013a, b; Busscher et al. 2012; Whitehouse et al. 2002). Infection that results from receiving a contaminated biomedical implant substantially affects the quality of life for the patient and represents a large burden to society by increasing the number of clinical procedures having to be undertaken in the healthcare system, thus increasing the costs of health care. For example, it has been estimated that antibiotic treatment and revisionary surgery for replacement of an infected total hip causes the primary procedure costs to triple to an average of $75,000 in the US (Lavernia et al. 2006; Bozic and Ries 2005). It has been found that infections that have arisen from orthopaedic surgery prolonged the hospital stay for patients by a median of 2 weeks each, approximately doubled the re-hospitalisation rates, and increased healthcare costs by more than 300 % (Whitehouse et al. 2002). Complications arise as a result of the the pathogenic bacteria developing resistance to the natural host defense mechanisms and antibiotics. This means that treatment of the infection whilst the implant is still resident in the patient is usually unsuccesful, leading to the necessity for surgical intervention (Olson et al. 2002; Davies 2003; Vasilev et al. 2009; Høiby et al. 2010). It appears that only a low dose of inoculum is required to result in the infection of an implant; in an animal model study, it was found that 100 colony-forming units (cfu) of *S. aureus* were sufficient to infect 95 % of the subcutaneous implants used in the study (Arciola et al. 2012; Campoccia et al. 2013b; Song et al. 2013). Most of the microorganisms that cause implant infections are present in the host flora, of which the most frequent are Staphylococci, Streptococci, Pseudomonas species and coliform bacteria (Campoccia et al. 2013a, b; Olson et al. 2002; Davies 2003).

In an effort to combat implant-associated bacterial infection, recent focus has been placed on the development of antibacterial surfaces (Campoccia et al. 2013a, b; McLean et al. 1993; Yoshinari et al. 2001; Wan et al. 2007a, b; Zhao et al. 2009; Vasilev et al. 2009; Rautray et al. 2010; Glinel et al. 2012; Hajipour et al. 2012; Hasan et al. 2013). With the current growth in the development of novel nanofabrication tools, modifying the chemical and physical characteristics of implant surfaces has been shown to control the molecular to micro-scale topological features of many substrate surfaces. This chapter will provide an insight into the current methods being used for the design of antibacterial surfaces on implant materials and a rationale for the applicability of the unique surfaces being produced.

6.2 Strategies Being Used for the Design of Antibacterial Implant Surfaces

6.2.1 *Overview*

Implantable biomaterials have long been used for restoring joint function, reduce pain or stabilise fractured bones (Pourbaix 1984; Cui and Luo 1999; Wan et al. 2007a, b; Zhao et al. 2009; Rautray et al. 2010; Minagar et al. 2012; Andani et al. 2014). Antibacterial implant materials need to be both antibacterial and biocompatible (Anselme et al. 2010; Zhao et al. 2009, 2014; Vasilev et al. 2009; Busscher et al. 2012; Campoccia et al. 2013a, b). In 1987, Anthony Gristina first introduced the concept of a "race for the surfaces" to describe the competition that exists between the integration of tissue cells onto the surface of an implant material and the adhesion of pathogenic bacteria (Gristina 1987; Gristina et al. 1988). Clearly, it is desirable for the host tissue to have the opportunity to establish itself over the surface of a biomedical implant in order that the implant surface is connected to the body, allowing further tissue generation to take place without the invasion of pathogenic bacteria. It is therefore not surprising that a variety of alternative approaches have been developed for the construction of biomaterials that exhibit antibacterial properties that can simultaneously support the host tissue integration and effectively deal with any bacteria coming into contact with the surface.

Here, the different approaches currently being used to reduce the vulnerability of medical devices to bacterial attachment will be discussed. The current designs for antibacterial biomaterials can be classified into two major groups, classified according to their mode of action. The first is antifouling surfaces, which have the ability to repel or prevent bacteria from adhering to their surface. The second is bactericidal surfaces, which have the ability to damage or kill any pathogenic bacteria coming into contact with the surface (Fig. 6.1).

6.2.2 *Antifouling Surfaces*

Bacterial adhesion to an implant surface occurs when bacteria are able to make contact with a surface that provides them with a favourable environment for changing from their planktonic state to that of being sessile; this is most often accompanied by the production of a bacterial biofilm (Olson et al. 2002; Anselme et al. 2010; Arciola et al. 2012; Foster et al. 2014). It is clear that an infection arising from the presence of pathogenic bacteria would not occur if the bacteria involved were

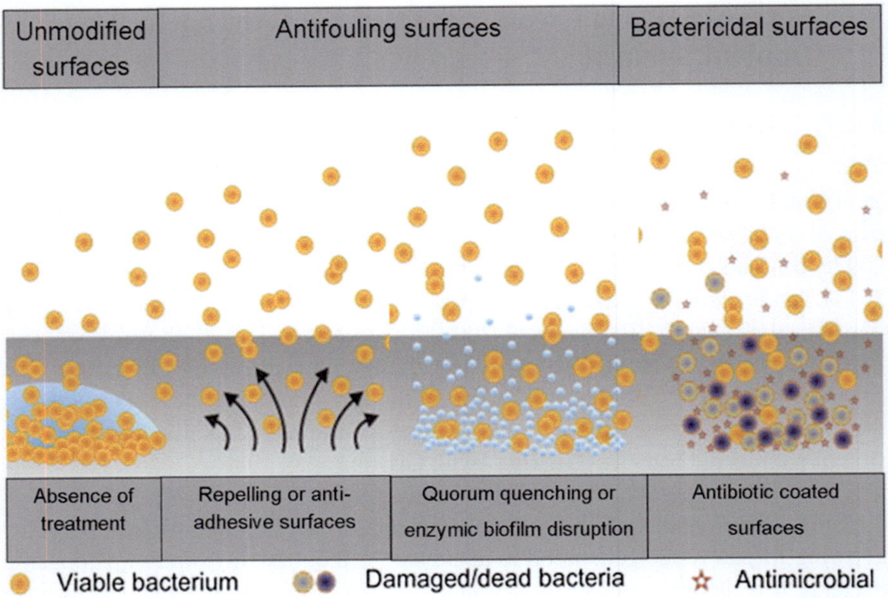

Fig. 6.1 Schematic representation of the different strategies currently being used in the design of antibacterial surfaces (Adapted from Campoccia et al. 2013a)

unable to initially colonise the medical device. The complex mechanisms associated with bacterial attachment have long been studied in order to gain an understanding into the methods by which metallic antibacterial surfaces can be designed such that this event can be prevented. A wide range of chemico-physical properties and functional groups on both the substrate and pathogen have been modified in order to modulate the attachment of these bacteria (Fusetani 2004, 2011; Hasan et al. 2013; Bazaka et al. 2011a, 2012; Webb et al. 2011; Crawford et al. 2012). More recently, surface architectures that contain specific surface porosity, roughness and geometry have been used to produce metallic biomaterial surfaces that are resistant to microbial colonisation (Meng et al. 2014; Anselme et al. 2010; Webb et al. 2011; Crawford et al. 2012; Bazaka et al. 2012).

Metallic biomaterial devices are often exposed to body fluids and a rich protein environment at the site of surgical implantation (Arciola et al. 2003; Campoccia et al. 2013a, b). It is known that a variety of host proteins promote bacterial attachment and the formation of biofilms. These are called microbial surface components recognizing adhesive matrix molecules, or MSCRAMMs, and these include collagen, fibrinogen, fibronectin, laminin, vitronectin, clumping factor A and B, bone-sialoprotein, elastin, IgG and other possible components (Lv et al. 2013; Montanaro et al. 2011; Patti et al. 1994; Arciola et al. 2012; Foster et al. 2014; Foster and Höök 1998; Hauck et al. 2006; Lambris et al. 2008). The biomaterial surfaces are therefore required to support the adsorption of host adhesins onto their surface to ensure the successful subsequent integration of tissue, whilst at the same time being able to repel the bacteria or be anti-adhesive towards them.

Chemical methods can be used to construct microbe-repellent surfaces by attaching antifouling molecules to the surfaces of implant materials (Neoh et al. 2012; Campoccia et al. 2013a, b). Common chemical modification approaches include rendering the surfaces superhydrophobic/superhydrophilic or highly hydrated or non-charged, each of these being unfavourable for bacterial adhesion under certain circumstances (Fig. 6.1) (Campoccia et al. 2013a, b). One of the most common coatings to render the surface hydrophilic is poly(ethylene glycol) (PEG). The inhibition mechanism of such surfaces is based on the dynamic motion and steric repulsion of hydrated polymer chains, which prevents bacterial attachment (Harris et al. 2004; Maddikeri et al. 2008). In addition, polycationic polymers exhibit antifouling effects have been used by directly coating or grafting them onto biomedical devices (Chua et al. 2008; Subbiahdoss et al. 2010; Shi et al. 2008; Hu et al. 2010; Siedenbiedel and Tiller 2012). Heparin coatings have also been shown to exhibit a high anti-adhesive effect for bacteria by increasing the hydrophilicity of the surfaces. The heparin forms a highly hydrated layer between the pathogens and the substrate (Ruggieri et al. 1987; Arciola et al. 1993). It was reported that heparin can inhibit the extent of *S. epidermidis* binding to fibronectin, thus preventing the subsequent colonisation of the surface (Arciola et al. 2003; Bustanji et al. 2003). Another approach, where quorum-sensing inhibitors (e.g. furanones and their derivatives) are incorporated onto biomedical device surfaces, was used to disrupt the processes responsible for the formation of a biofilm (Fig. 6.1) (Fusetani 2004, 2011). This approach, however, has significant drawbacks in terms of the long-term stability of the coating and the possible cytotoxicity of these additives in biomedical applications. Current approaches use surface topography as the factor by which the degree of bacterial adhesion and subsequent biofilm formation can be controlled or prevented. Techniques such as this represent a more robust method for creating surfaces that repel or control the extent of microbe attachment (Webb et al. 2011, 2014; Hasan et al. 2013; Grinthal and Aizenberg 2014; Bazaka et al. 2012; Crawford et al. 2012). For example, superhydrophobic surfaces have been shown to exhibit antifouling characteristics and can be obtained by physically modifying the micro- and nanostructures of biomaterial surfaces by mimicking natural surface structures such as that of the lotus leaf (Truong et al. 2012; Fadeeva et al. 2011; Crick et al. 2011). By tailoring the precise and specific surface topographical parameters, these surfaces have shown promising results in their ability to limit the initial adhesion of pathogenic bacteria (Webb et al. 2011, 2014; Hasan et al. 2013).

6.2.3 Bactericidal Surfaces

Another common approach in the prevention of biofilms on biomedical devices is the utilization of bioactive antibacterial agents that act by contact killing the bacteria. These techniques involve coating the substrate with various immobilized antimicrobial substances such as antibacterial peptides (Brouwer et al. 2011; McCloskey et al. 2014; Salwiczek et al. 2014), quaternary amines (Mei et al. 2012; Schaer et al. 2012), nitric oxide (Fox et al. 2010; Nablo et al. 2005) or antibacterial metals (silver, zinc, cobalt, aluminium and copper) (Kawashita et al. 2000; McLean et al. 1993; Heidenau et al. 2005; Lemire et al. 2013; Stafford et al. 2013; Prantl et al. 2010; Wan et al. 2007a). These substances are not released from the substrate, and as such they directly interact with any pathogenic bacteria coming in contact with the surface (Williams and Worley 2000). Bioactive antibacterial coatings have been used extensively in applications that require the surface to be self-sterilizing over extended periods (Williams and Worley 2000; Campoccia et al. 2013a).

Silver and its derivatives are some of the earliest bactericidal agents that have been largely applied in a wide range of applications (Zhao et al. 2009; Kawashita et al. 2000; McLean et al. 1993; Nomiya et al. 1997; Dueland et al. 1982; Richards 1981; Bayston et al. 2010). Other metals that have also been reported to exhibit bactericidal effects, mostly in their composite form, include zinc, cobalt, aluminium and copper (Heidenau et al. 2005; Lemire et al. 2013; Samanovic et al. 2012; Stafford et al. 2013; Hoene et al. 2013; Prantl et al. 2010; Shirai et al. 2009; Wan et al. 2007a, b). The use of antimicrobial metals is, however, often associated with certain degree of cytotoxicity. This can have an impact on the host cell response, leading to the loss of cell viability and the failure of tissue integration (Heidenau et al. 2005; Hoene et al. 2013; Paasche et al. 2011; Vasilev et al. 2009). This occurs mainly as a result of corrosion of the metal in the physiological environment, which causes the release of metal ions at relatively high concentrations, leading to local toxicity and occasionally metal accumulation in the target organs (Campoccia et al. 2013b; Lemire et al. 2013; Vasilev et al. 2009). The mechanisms responsible for the antibacterial activity of metals and metal ions is not fully understood. Gordon et al. suggested that silver interacts with thiol groups, causing the inactivation of critical enzymes in the respiratory chain and the induction of hydroxyl radicals (Gordon et al. 2010).

Another emerging strategy for the manufacture of antimicrobial surfaces is the incorporation of biocide-releasing surfaces such as those containing nanoparticles. The extent of the bactericidal effect of these surfaces depends on the size, shape, concentration and chemical composition of the nanoparticles (Cui et al. 2012b; Zhang et al. 2013; Hajipour et al. 2012). While the exact mechanisms of the antimicrobial activity are also not fully understood, most nanoparticles are seen to generate reactive oxygen species and damage the cell membranes (Cui et al. 2012b; Zhang et al. 2013; Hajipour et al. 2012). For example, gold nanoparticles exhibit bactericidal effects against *E. coli* by inhibiting ATP synthase activity, followed by the inhibition of the ribosome sub-unit in tRNA binding (Cui et al. 2012b). There is still a lack of knowledge on the toxicology of nanoparticles, with most of the available data being inconsistent and largely non-reproducible (Campoccia et al. 2013a; Yildirimer et al. 2011). The negative impact of nanoparticles in biomedical applications includes the induction of apoptosis, introduction of toxic effects to the genome and the possible translocation of nanoparticles to distant tissues and organs, with an associated risk of systemic effects (Yildirimer et al. 2011; Campoccia et al. 2013a).

The major problem, however, is that biofilms display an increased tolerance towards antimicrobial agents, which substantially restricts our ability to treat biofilm-related infections in clinical settings. While the increased resilience of biofilms towards antibiotics is multifactorial, this resistance can be attributed to the presence of persistent bacteria, those that can enter into a specific phenotype state that allows them to survive in the presence of 1000 times the minimum inhibitory concentration of bactericidal antibiotics (Olson et al. 2002; Davies 2003). Persistent cells have recently been the subject of increased investigation with a view to limiting their biofilm-associated antibiotic tolerance. The current strategy for preventing the formation of biofilms has been to develop ways by which the initial bacterial adhesion step can be inhibited, which will subsequently limit the growth of the biofilm (Hasan et al. 2013; Fusetani 2004).

Recently, the effects of surface topography on the attachment responses of bacterial and mammalian cells has been under investigation in an effort to obtain an insight into the competition that takes place when bacteria and host tissue compete for attachment on a substrate surface (Hasan et al. 2013). When trying to prevent biofilms from forming on medical implant surfaces, a common approach is to develop a surface structure that can both physically inhibit the growth of bacteria, but at the same time promote tissue integration. Nano-structured surfaces have shown numerous promising results. Interesting reports have demonstrated that surfaces containing nanopillar arrays that mimic the structure of dragonfly wings can exhibit a bactericidal effect to not only Gram positive and Gram negative bacteria, but also their spores (Hasan et al. 2013; Ivanova et al. 2013; Bazaka et al. 2012). Recent research has also shown that mammalian cells are biocompatible with these high aspect ratio structured surfaces that contain complex geometries, in which the cells appeared to be able to maintain their viability, adhesion to the surface and their subsequent cellular activities (Kim et al. 2007; Robinson et al. 2012; Elnathan et al. 2014; Jahed et al. 2014). For the reasons previously described, the ability to design surfaces that possess antimicrobial properties without the need for the surface to contain antibiotics or chemical additives represents a significant step forward in developing implant materials that are less likely to be the cause of post-operative infections.

6.3 Fabrication Techniques

Since the advent of micro- and nano-technology, a number of fabrication techniques have been developed that can be used to modify the surface properties of metallic implants on a molecular-, nano- and micro-scale. Antibacterial metallic surfaces can be fabricated via two principal methods: chemical and physical modification (Jeon et al. 2014; Lv and Feng 2006; Yoshinari et al. 2001; Vasilev et al. 2009;

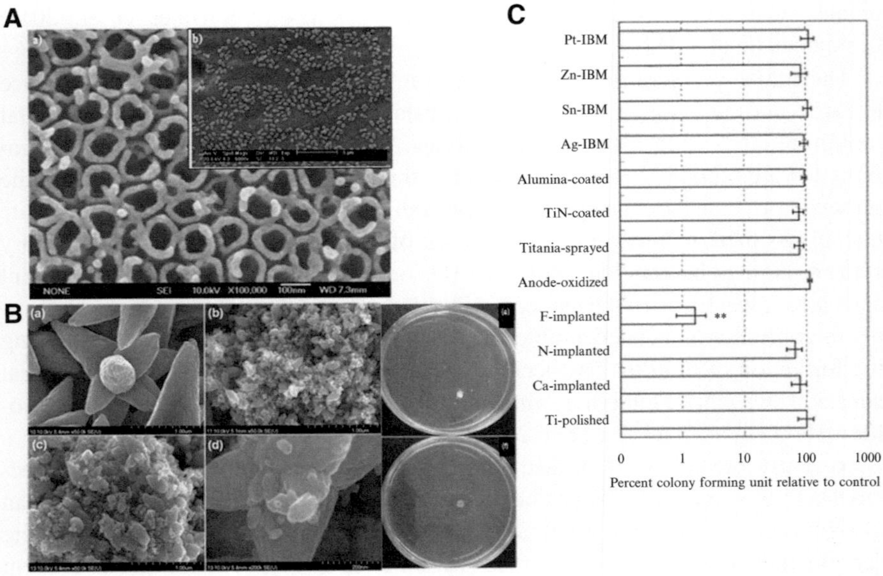

Fig. 6.2 Representative antibacterialmetallic surfaces fabricated via various chemical fabrication techniques. (**A**) (*a*) SEM images of the *top-view* of TiO_2 nanotube surfaces; the nanotubes are approximately 40–97 nm in diameter and 300 nm in length. (*b*) (inset) the changed morphologies of the surface after the adhesion patterns of *Streptococcus mutans*, cultured for 48 h (Adapted with permission from Cui et al. 2012a, b). (**B**) FE-SEM images of (*a*) pure ZnO, (*b*) TiO_2/ZnO, (*c*) Ag/TiO_2/ZnO particles, and (*d*) high magnification of (*c*). (*e*) and (*f*) represent the zones of inhibition tests for (*top*) TiO_2/ZnO and (*bottom*) Ag/TiO_2/ZnO composite surfaces towards *E. coli* (Adapted with permission from Pant et al. 2013). (**C**) Colony-forming units relative to control against *P. gingivalis* on 1 cm^2 plates for 48 h against different ion-implanted Ti surfaces (Adapted with permission from Yoshinari et al. 2001)

Hasan et al. 2013; Salwiczek et al. 2014). Techniques such as surface chemical functionalization, chemical vapour deposition, anodic oxidation, hydrothermal treatment and ion implantation involve chemical reactions occurring at the surface to modify the surface properties. Techniques such as physical vapour deposition, layer-by-layer coating, and sol-gel coating utilise physical adsorption on implant surfaces without altering the surface chemistry.

6.3.1 Chemical Modification

This section highlights the techniques that employ processes such as wet chemistry and high energy sources to alter the surface characteristics of metal surfaces. Here, we present a general overview of the chemical modification techniques that can be used to produce antibacterial metallic surfaces in Figs. 6.2 and 6.3.

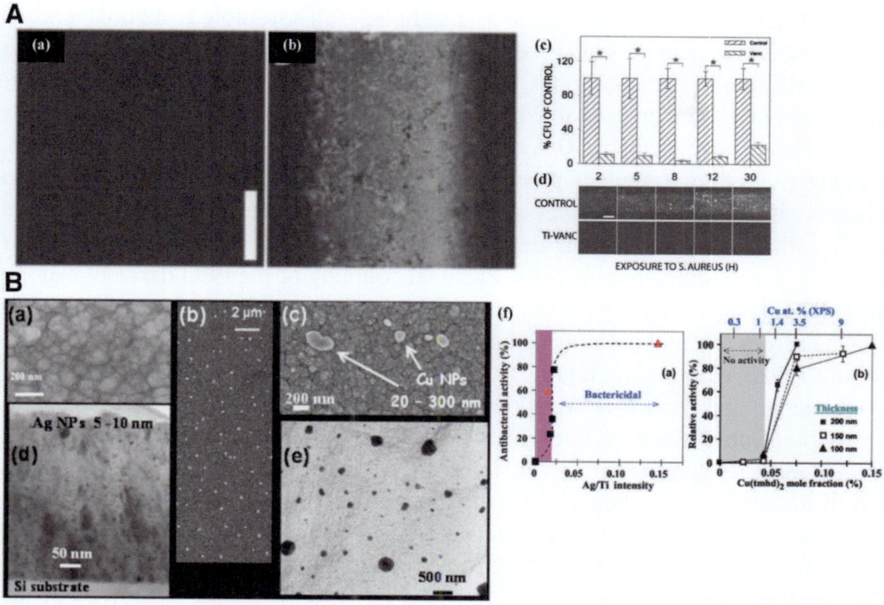

Fig. 6.3 (**A**) Confocal microscopic images of (*a*) a control Ti rod and (*b*) Ti rods covalently linked with vancomycin (vanc); (*c*) The antibacterial efficiency of the control and Ti-vanc rods towards *S. aureus* incubated at a time period from 0 to 30 h. The Ti-Vanc rods showed fewer adherent colonies at all times than control rods. (*d*) Represents the fluorescent stains for the viable cells on the control Ti surface (Adapted with permission from Antoci et al. 2007). (**B**) SEM images of (*a*) Ag-TiO_2 and (*b*, *c*) Cu-TiO_2 films. (*d*) TEM cross section micrograph of Ag-TiO_2 and (*e*) planar view of Cu-TiO_2 coatings show the metal nanoparticle distribution in the TiO_2 matrix through chemical vapour deposition (CVD). (*f*) The influence of Ag (*left*) and Cu (*right*) content of M-TiO_2 nanocomposite coatings on the antibacterial behaviour against *S. aureus*. The *coloured zone* corresponds to inactive surfaces according to the JIZ test (Adapted with permission from Maury et al. 2014)

6.3.1.1 Chemical Functionalisation

Chemical functionalization of various metallic surfaces has been one of the most commonly used techniques for preparing antibacterial surfaces (Zobrist et al. 2011; Ogaki et al. 2010; Gerberich and Bhatia 2013). This fabrication technique is suitable for designing antifouling and bactericidal surfaces that can kill bacteria on contact. The functionalisation of inert metallic surfaces is, however, a challenge. To overcome this problem, there are two strategies used for functionalising surfaces such that they contain antifouling or bactericidal agents (Yuan et al. 2011; Gadenne et al. 2013; Antoci et al. 2007; Godoy-Gallardo et al. 2014; Holmberg et al. 2013; Chen et al. 2013): These are: (1) developing adhesion coatings, such as self-assembled monolayers (SAM); or (2) functionalising metallic surfaces with active groups, such as thiol (-SH), hydroxyl (-OH), amine (-NH_2) or carboxylic groups (-COOH).

The first strategy is to coat the surface with self-assembled monolayers (SAM) to act as an adhesion layer for the immobilisation of antibacterial agents (Yuan et al.

2011; Gadenne et al. 2013). For example, self-assembled monolayers of aminoundecyltrimethoxysilane were used as an adhesion layer on Ti substrates to covalently bond polysaccharides extracted from *Ulva rotundata* and *Ulva compressa* seaweed (Gadenne et al. 2013). Adhesion of *Pseudomonas aeruginosa* cells was shown to be reduced by 90 % on these surfaces compared to the control surface. Conjugation of poly(ethylene glycol) (PEG) brushes on stainless steel was used to couple with hydrolytic enzymes such as lysozyme (Yuan et al. 2011), creating a surface that exhibited a high degree of antifouling and bactericidal behaviour towards *E. coli* and *S. aureus* cells. The dual coupling of the PEG derivative along with lysozyme served a dual function; displaying antifouling properties towards bovine serum albumin protein and antibacterial behaviour towards the specific bacterial strains (Yuan et al. 2011).

A second approach is to functionalise surface with reactive groups, then covalently bind antibiotics or antimicrobial peptides (AMP) to the surface (Antoci et al. 2007; Godoy-Gallardo et al. 2014; Holmberg et al. 2013; Chen et al. 2013). A number of studies have been reported for the successful tethering of antibiotics to metallic implant surfaces. For example, non-reactive Ti surfaces were chemically modified to carry amine groups that reacted with vancomycin (Antoci et al. 2007). This method was shown to be able to preserve the antibacterial properties of the modified Ti surfaces, while preventing the release of antibiotics into the surrounding body fluid. In other studies, AMP were conjugated onto metallic surfaces. For example, Ti surfaces were chemically modified to contain hydroxyl groups, onto which hydrophilic poly (ethylene glycol) was conjugated as a spacer between the AMP and the Ti surface (Gabriel et al. 2006).

6.3.1.2 Chemical Vapour Deposition

Chemical vapour deposition (CVD) is yet another chemical process that has been used to fabricate an antibacterial layer over substrate surfaces on a large scale (Wilkinson et al. 2013; Bazaka et al. 2010, 2011a, b, 2012; Dastjerdi and Montazer 2010; Maury et al. 2014; Varghese et al. 2013). In a deposition process, the substrate is exposed to antibacterial monomer precursors, which react with the substratum under plasma condition to synthesise antibacterial surfaces. These processes have been used to fabricate both inorganic and organic antibacterial coatings.

Inorganic coatings on metallic implant surfaces such as silver, TiO_2 and other photocatalytic metal oxides was performed using CVD (Dastjerdi and Montazer 2010; Maury et al. 2014; Varghese et al. 2013). Composite TiO_2 films, coupled with other metallic ions such as Ag^+ or Cu^{2+}, have also been synthesized using CVD (Maury et al. 2014). TiO_2 nano-composite films coupled with Ag have been shown to exhibit strong bactericidal efficiency against *S. aureus*, where a minimum film thickness of 100 nm has been found to be essential for rendering the film antibacterial in nature (Maury et al. 2014). Flame-assisted chemical vapour deposition has also been recently employed to fabricate silver-silica coatings on glass surfaces, the antibacterial efficiencies of which have been tested using *E. coli*, *S. aureus* and *P.*

aeruginosa (Varghese et al. 2013). A plasma deposition method has also been used to deposit Cu ions on titaniumalloy surfaces. This combination of Cu ions embedded onto Ti_6Al_4V surfaces has proven to be an effective method for preparing an antibacterial surface, since the amount of copper ions released from the surface resulted in killing ~99 % of the bacterial cells (Hempel et al. 2014).

For organic coatings to be prepared using CVD, monomers of antibacterial agents need to be used as precursors for the formation of antibacterial polymer films on a substrate (Bazaka et al. 2010, 2011a, b; Pegalajar-Jurado et al. 2014). For example, substrata have been encapsulated with polyterpenol thin films prepared from antibacterial terpinen-4-ol monomers using radio frequency plasma enhanced chemical vapour deposition (Bazaka et al. 2010, 2011b). These polyterpenol films were found to be effective against *P. aeruginosa* and *S. aureus* cells due to the preservation of original terpinen-4-ol molecules in the resulting film structure (Bazaka et al. 2010, 2011b). In another study, plasma polymerised thin films produced from antibacterial 1,8-cineole monomers were shown to inhibit the proliferation of *E. coli* and *S. aureus* (Pegalajar-Jurado et al. 2014).

6.3.1.3 Anodic Oxidation

Electrochemical anodic oxidation techniques have been used to fabricate highly ordered nanoporous nanotubes on metal surfaces (Cui et al. 2012a; Li et al. 2013; Yue et al. 2014; Minagar et al. 2012; Visai et al. 2011). In particular, TiO_2 nanotubes have been recognized as promising biomaterials, with proven biocompatibility, thermal stability and corrosion resistance. Enhancement of the antibacterial properties of TiO_2 nanotubes can be achieved through the use of UV illumination or the incorporation of antibiotic loadings (Cipriano et al. 2014; Cui et al. 2012a; Çalışkan et al. 2014; Chennell et al. 2013). For example, TiO_2 nanotube layers have been designed to become photocatalytic, which can disrupt the viability of *Streptococcus mutans* (Cui et al. 2012a), whereas the incorporation of gentamycin inside the nanotube patterns was found to improve their antibacterial properties (Çalışkan et al. 2014).

6.3.1.4 Hydrothermal Treatment

Hydrothermal synthesis is an environmental friendly process for synthesising antibacterial surfaces because the synthesis reaction is completely contained in aqueous solutions inside a closed system, making use of acidic or basic solutions as the reaction medium (Huo et al. 2013; Pant et al. 2013; Hebeish et al. 2013; Li et al. 2014; Wong et al. 2011). In such processes, the operating temperature is held above the boiling point of water to autogenously produce a saturated vapour pressure. As with the anodic oxidation technique, this technique is only suitable for the preparation of antibacterial inorganic coatings, such as silver, TiO_2, and other photocalytic materials (Huo et al. 2013; Pant et al. 2013; Hebeish et al. 2013; Li et al. 2014).

For example, a one-step hydrothermal synthesis has been used to fabricate ZnO and silver nanoparticles coupled with a passive TiO_2 layer to improve the antibacterial efficiency of photocatalytic TiO_2 layers (Pant et al. 2013). In addition, it has been reported that Ti nanowires doped with silver exhibited commendable antimicrobial efficiency towards various *Pseudomonas* species (Hebeish et al. 2013; Li et al. 2014).

6.3.1.5 Ion Implantation

Ion implantation has been used as a method for doping antibacterial inorganic ions onto the surface of biomedical implants under the immersion of plasma (Cui and Luo 1999; Rautray et al. 2010; Lu et al. 2012; Yoshinari et al. 2001; Zhao et al. 2009). Ti and Ti alloys modified with ion implantation (F^+) were shown to inhibit the growth of both *Porphyromonas gingivalis* and *Actinobacillus actinomycetemcomitan* (Yoshinari et al. 2001; Rautray et al. 2010). In addition, it was shown that F+implanted surfaces did not inhibit the proliferation of fibroblast L929-cells (Yoshinari et al. 2001). Silver and copper ion-doped Ti surfaces prepared using ion implantation processes were shown to reduce the viability of *S. aureus* cells coming into contact with the surface (Wan et al. 2007a).

6.3.2 Physical Modification

Physical modification techniques coat metallic substrates through physical adsorption without altering the surface chemistry. Here, we present a general overview of the physical modification techniques that can be used for the preparation of antibacterial metallic surfaces (Fig. 6.4).

6.3.2.1 Physical Vapour Deposition (PVD)

Physical vapour deposition (PVD), due to its environmentally friendly characteristics, convenience and precision in deposition, has become one of the commonly used techniques in preparing uniform surface coatings (Zaborowska et al. 2014; Percival et al. 2005; Ip et al. 2006; Ivanova et al. 2011; Chang et al. 2013; Trivedi et al. 2014). A number of reports have shown that physical vapour deposition has also been used to deposit uniform coatings on titanium and its alloys. This technique fabricates an antibacterial surface by employing a twin-gun magnetron sputtering system to synthesize uniform coatings of silver and zirconium oxide (Zaborowska et al. 2014; Percival et al. 2005; Ip et al. 2006; Ivanova et al. 2011; Chang et al. 2013). This process is also commonly used to deposit silver coatings on titanium surfaces, since the silver coating provides a biocompatible surface and also helps to reduce the attachment and viability of bacteria (Zaborowska et al. 2014; Percival et al. 2005; Ip

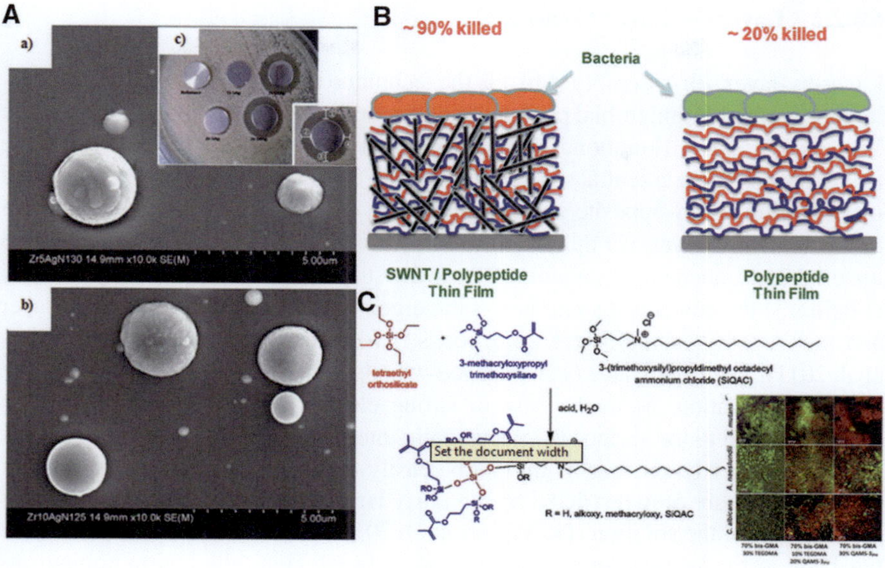

Fig. 6.4 Representative antibacterial metallic surface fabricated via various physical fabrication techniques. (**A**) FE-SEM images of the (*a*) TiAgN and (*b*) ZrAgN coatings fabricated through physical vapour deposition (PVD). *Scale bars* indicate 5 μm. *Inset* image (*c*) represents the clear zone of inhibitions of the coated samples against *S. mutans* after 18 h incubation (Adapted with kind permission from Kang and Lim 2014). (**B**) A schematic illustration of the antimicrobial property of the layer-by-layer assembly of single wall carbon nanotubes along with the polyelectrolytes poly(L-lysine) and poly(L-glutamic acid) (Adapted from Aslan et al. 2012). (**C**) Reaction scheme for producing quaternary ammonium silanes (QAMS) using tetra-alkoxysilane as attaching unit. Macromonomers with monofunctional (QAMS-1), bifunctional (QAMS-2) or trifunctional methacryloxy functionalities (QAMS-3) have been fabricated based on the molar ratio of the two trialkoxysilanes through the process of sol-gel coating. Inset image represents the antimicrobial activity of the polymerized resin through the confocal laser scattering microscopy images of 48 h microbial biofilms of *S. mutans* (*top*), *A. naeslundii* (*middle*) and *C. albicans* (*bottom*) respectively (Adapted with permission from Gong et al. 2012)

et al. 2006; Ivanova et al. 2011; Chang et al. 2013). Additionally, this system offers almost no cytotoxicity issues towards mammalian cells.

Another slight variation of this technique, arc-ion plating (AIP), has also been used to fabricate TiAgN and ZrAgN alloys coatings (Kang and Lim 2014). These coatings serve as efficient antibacterial coatings by reducing the adhesion and viability of *S. mutans* (Kang and Lim 2014). AIP is a widely used technique in the biomedical industry due to its advantages, including its ability to produce a dense metal vapour, high ionization efficiency and high deposition rate (Joo et al. 2009). Plasma nitriding of stainless steel surfaces with Ag^{+} ions has also resulted in the synthesis of highly efficient antimicrobial surfaces, which have been reported to eliminate almost all (~97 %) of the inoculating bacterial cells of *E. coli* and *S. epidermidis* in a 6 h time span by the method of contact killing (Dong et al. 2011).

6.3.2.2 Layer-by-Layer Coatings

Layer-by-layer self-assembly (LbL) is the technique that has been used to encapsulate antibiotics, antimicrobial peptides and nanomaterials onto metallic surfaces (De Villiers et al. 2011; Hammond 1999, 2004; Decher 1997; Linford et al. 1998). The technique has some essential steps, as follows: (1) A charged substrate is submerged in a solution of an oppositely-charged colloid, designed to assist the adsorbtion of the first monolayer, (2) this is followed by a washing cycle to remove unbound material and impede the contamination of the surface with the oppositely-charged colloid, (3) the coated substrate is re-submerged to finally adsorb the second layer and thus a series of multi-layered deposits is formed in this manner (De Villiers et al. 2011). There also have been instances where no washing step is required in the sample preparation, as in the case of strong electrolytes, where the polymer is strongly bound to the surface by electrostatic interactions (Linford et al. 1998).

Feature sizes lesser than 1 μm can be easily obtained, since the limitations of this technique are only restricted by the fairly large dimensions of the bound macromolecules in the solution (De Villiers et al. 2011). A recent use of this technique has been in the fabrication of antibacterial surfaces. Here, the antimicrobial efficiency of single-walled carbon nanotubes (SWNT) that were layer-by-layer assembled with poly (L-lysine) and poly (L-glutamic acid) were well studied for their resistance to the attachment of the bacterial strains of *E. coli* and S. *epidermidis* (Aslan et al. 2012, 2013). This assembly of multilayer films has the ability of reducing the proliferation rates of the bacterial cells by up to 90 % (Aslan et al. 2012, 2013). With LbL technology, multi-faceted surfaces with altered shapes can be conveniently coated with conformal ultra-thin films (Hammond 2004; Decher 1997). Layer by layer assembly has also been used to fabricate a cross-linked polymeric thin film using a polycation, N, N-dodecyl, methyl-polyethylenimine and a polyanion, poly (acrylic acid). Surfaces containing this combination have been reported to be highly effective against the commonly found infecting strains of *E. coli* and *S. aureus* (Wong et al. 2010). This film exhibited considerable antibacterial effectiveness by causing cell lysis when the cells came in contact with the surface of the polymeric film.

6.3.2.3 Sol-Gel Coating

Sol-gel coating processes involve the conversion of monomers into antibacterial colloidal solutions (sols) that acts as precursors for an integrated network (gel) of the coatings (Chun et al. 2007; Rivero et al. 2011; Talebian et al. 2014; Visai et al. 2011). Sol-gel TiO_2 coatings on stainless steel orthodontic wires have been shown to reduce the viability of *Streptococcus mutans* and *Porphyromonas gingivalis* (Chun et al. 2007). A novel antibacterial coating composed of an organic-inorganic hybrid matrix of tetraorthosilicate and a polyelectrolyte was successfully doped with Ag nanoparticles through sol-gel processes (Rivero et al. 2011). This specific coating has ability to eliminate the growth of *Lactobacillus plantarum*. Nickel oxide nanoparticles that were immobilised in a sol-gel coating were also found to exhibit a high degree of antibacterial activity against *S. aureus* and *E. coli* cells (Talebian et al. 2014).

6.4 Future Perspectives

In general, this chapter provided an overview of the current approaches and techniques that have been developed in the design of antibacterialmetallic implant surfaces. Metallic implant surfaces can be modified to become either antifouling or bactericidal in nature. To achieve this, there are a number of chemical or physical modification techniques that have been developed for the fabrication of antibacterial metallic implant surfaces. In the last few decades, the surface chemical characteristics of a surface have been modified such that the resulting surface is not only antibacterial, but also biocompatible. Investigation of the cytotoxicity and biocompatibility of such modified materials requires significant resources. One of the currently emerging approaches for determining the cytotoxicity and biocompatibility of a surface is to develop surface micro- and nano-architectures that generate repulsive forces towards bacteria (Ivanova et al. 2012, 2013; Hasan et al. 2013). These techniques have proven to be a good starting point for the innovative design of metallic biomaterials, and represent an alternative approach to the methods that have traditionally been adopted. To date, only a limited number of studies have been performed to address these issues (Ivanova et al. 2012, 2013; Fadeeva et al. 2011). These recently developed methods for modifying the nanotopography of surfaces may prove to be very useful techniques for the fabrication of a new generation of antifouling or bactericidal biomaterials.

References

Andani MT, Shayesteh Moghaddam N, Haberland C, Dean D, Miller MJ, Elahinia M (2014) Metals for bone implants. Part 1. Powder metallurgy and implant rendering. Acta Biomater 10(10):4058–4070. doi:10.1016/j.actbio.2014.06.025

Anselme K, Davidson P, Popa AM, Giazzon M, Liley M, Ploux L (2010) The interaction of cells and bacteria with surfaces structured at the nanometre scale. Acta Biomater 6(10):3824–3846. doi:10.1016/j.actbio.2010.04.001

Antoci V, King SB, Jose B, Parvizi J, Zeiger AR, Wickstrom E, Freeman TA, Composto RJ, Ducheyne P, Shapiro IM (2007) Vancomycin covalently bonded to titanium alloy prevents bacterial colonization. J Orthop Res 25(7):858–866. doi:10.1002/jor.20348

Arciola CR, Radin L, Alvergna P, Cenni E, Pizzoferrato A (1993) Heparin surface treatment of poly(methylmethacrylate) alters adhesion of a Staphylococcus aureus strain: utility of bacterial fatty acid analysis. Biomaterials 14(15):1161–1164. doi:10.1016/0142-9612(93)90161-T

Arciola CR, Bustanji Y, Conti M, Campoccia D, Baldassarri L, Samorì B, Montanaro L (2003) Staphylococcus epidermidis-fibronectin binding and its inhibition by heparin. Biomaterials 24(18):3013–3019. doi:10.1016/S0142-9612(03)00133-9

Arciola CR, Campoccia D, Speziale P, Montanaro L, Costerton JW (2012) Biofilm formation in *Staphylococcus* implant infections. A review of molecular mechanisms and implications for biofilm-resistant materials. Biomaterials 33(26):5967–5982. doi:10.1016/j.biomaterials.2012.05.031

Aslan S, Deneufchatel M, Hashmi S, Li N, Pfefferle LD, Elimelech M, Pauthe E, Van Tassel PR (2012) Carbon nanotube-based antimicrobial biomaterials formed via layer-by-layer assembly with polypeptides. J Colloid Interface Sci 388(1):268–273. doi:10.1016/j.jcis.2012.08.025

Aslan S, Määttä J, Haznedaroglu BZ, Goodman JP, Pfefferle LD, Elimelech M, Pauthe E, Sammalkorpi M, Van Tassel PR (2013) Carbon nanotube bundling: influence on layer-by-layer assembly and antimicrobial activity. Soft Matter 9(7):2136–2144. doi:10.1002/adma.201001215

Bayston R, Vera L, Mills A, Ashraf W, Stevenson O, Howdle SM (2010) In vitro antimicrobial activity of silver-processed catheters for neurosurgery. J Antimicrob Chemother 65(2):258–265. doi:10.1093/jac/dkp420

Bazaka K, Jacob MV, Truong VK, Wang F, Pushpamali WAA, Wang JY, Ellis AV, Berndt CC, Crawford RJ, Ivanova EP (2010) Plasma-enhanced synthesis of bioactive polymeric coatings from monoterpene alcohols: a combined experimental and theoretical study. Biomacromolecules 11(8):2016–2026. doi:10.1021/bm100369n

Bazaka K, Jacob MV, Crawford RJ, Ivanova EP (2011a) Plasma-assisted surface modification of organic biopolymers to prevent bacterial attachment. Acta Biomater 7(5):2015–2028. doi:10.1016/j.actbio.2010.12.024

Bazaka K, Jacob MV, Truong VK, Crawford RJ, Ivanova EP (2011b) The effect of polyterpenol thin film surfaces on bacterial viability and adhesion. Polymer 3(1):388–404. doi:10.3390/polym3010388

Bazaka K, Jacob MV, Crawford RJ, Ivanova EP (2012) Efficient surface modification of biomaterial to prevent biofilm formation and the attachment of microorganisms. Appl Microbiol Biotechnol 95(2):299–311. doi:10.1007/s00253-012-4144-7

Bozic KJ, Ries MD (2005) The impact of infection after total hip arthroplasty on hospital and surgeon resource utilization. J Bone Joint Surg Ser A 87(8):1746–1751. doi:10.2106/JBJS.D.02937

Brouwer CPJM, Rahman M, Welling MM (2011) Discovery and development of a synthetic peptide derived from lactoferrin for clinical use. Peptides 32(9):1953–1963. doi:10.1016/j.peptides.2011.07.017

Busscher HJ, Van Der Mei HC, Subbiahdoss G, Jutte PC, Van Den Dungen JJAM, Zaat SAJ, Schultz MJ, Grainger DW (2012) Biomaterial-associated infection: locating the finish line in the race for the surface. Sci Transl Med 4(153):153rv110. doi:10.1126/scitranslmed.3004528

Bustanji Y, Arciola CR, Conti M, Mandello E, Montanaro L, Samorí B (2003) Dynamics of the interaction between a fibronectin molecule and a living bacterium under mechanical force. Proc Natl Acad Sci U S A 100(23):13292–13297. doi:10.1073/pnas.1735343100

Çalışkan N, Bayram C, Erdal E, Karahaliloğlu Z, Denkbaş EB (2014) Titania nanotubes with adjustable dimensions for drug reservoir sites and enhanced cell adhesion. Mater Sci Eng C 35:100–105. doi:10.1016/j.msec.2013.10.033

Campoccia D, Montanaro L, Arciola CR (2013a) A review of the biomaterials technologies for infection-resistant surfaces. Biomaterials 34(34):8533–8554. doi:10.1016/j.biomaterials.2013.07.089

Campoccia D, Montanaro L, Arciola CR (2013b) A review of the clinical implications of anti-infective biomaterials and infection-resistant surfaces. Biomaterials 34(33):8018–8029. doi:10.1016/j.biomaterials.2013.07.048

Chang Y-Y, Huang H-L, Chen Y-C, Weng J-C, Lai C-H (2013) Characterization and antibacterial performance of ZrNO–Ag coatings. Surf Coat Technol 231:224–228. doi:10.1016/j.surfcoat.2012.05.084

Chen X, Sevilla P, Aparicio C (2013) Surface biofunctionalization by covalent co-immobilization of oligopeptides. Colloids Surf B: Biointerfaces 107:189–197. doi:10.1016/j.colsurfb.2013.02.005

Chennell P, Feschet-Chassot E, Devers T, Awitor KO, Descamps S, Sautou V (2013) In vitro evaluation of TiO_2 nanotubes as cefuroxime carriers on orthopaedic implants for the prevention of periprosthetic joint infections. Int J Pharm 455(1–2):298–305. doi:10.1016/j.ijpharm.2013.07.014

Chua PH, Neoh KG, Kang ET, Wang W (2008) Surface functionalization of titanium with hyaluronic acid/chitosan polyelectrolyte multilayers and RGD for promoting osteoblast functions and inhibiting bacterial adhesion. Biomaterials 29(10):1412–1421. doi:10.1016/j.biomaterials.2007.12.019

Chun MJ, Shim E, Kho EH, Park KJ, Jung J, Kim JM, Kim B, Lee KH, Cho DL, Bai DH, Lee SI, Hwang HS, Ohk SH (2007) Surface modification of orthodontic wires with photocatalytic titanium oxide for its antiadherent and antibacterial properties. Angle Orthod 77(3):483–488. doi:10.2319/0003-3219(2007)077[0483:SMOOWW]2.0.CO;2

Cipriano AF, Miller C, Liu H (2014) Anodic growth and biomedical applications of TiO_2 nanotubes. J Biomed Nanotechnol 10(10):2977–3003. doi:10.1166/jbn.2014.1927

Crawford RJ, Webb HK, Truong VK, Hasan J, Ivanova EP (2012) Surface topographical factors influencing bacterial attachment. Adv Colloid Interf Sci 179–182:142–149. doi:10.1016/j.cis.2012.06.015

Crick CR, Ismail S, Pratten J, Parkin IP (2011) An investigation into bacterial attachment to an elastomeric superhydrophobic surface prepared via aerosol assisted deposition. Thin Solid Films 519(11):3722–3727. doi:10.1016/j.tsf.2011.01.282

Cui FZ, Luo ZS (1999) Biomaterials modification by ion-beam processing. Surf Coat Technol 112(1–3):278–285. doi:10.1016/S0257-8972(98)00763-4

Cui C, Gao X, Qi Y, Liu S, Sun J (2012a) Microstructure and antibacterial property of in situ TiO_2 nanotube layers/titanium biocomposites. J Mech Behav Biomed Mater 8:178–183. doi:10.1016/j.jmbbm.2012.01.004

Cui Y, Zhao Y, Tian Y, Zhang W, Lü X, Jiang X (2012b) The molecular mechanism of action of bactericidal gold nanoparticles on Escherichia coli. Biomaterials 33(7):2327–2333. doi:10.1016/j.biomaterials.2011.11.057

Dastjerdi R, Montazer M (2010) A review on the application of inorganic nano-structured materials in the modification of textiles: focus on anti-microbial properties. Colloids Surf B: Biointerfaces 79(1):5–18. doi:10.1016/j.colsurfb.2010.03.029

Davies D (2003) Understanding biofilm resistance to antibacterial agents. Nat Rev Drug Discov 2(2):114–122. doi:10.1038/nrd1008

De Villiers MM, Otto DP, Strydom SJ, Lvov YM (2011) Introduction to nanocoatings produced by layer-by-layer (LbL) self-assembly. Adv Drug Deliv Rev 63(9):701–715. doi:10.1016/j.addr.2011.05.011

Decher G (1997) Fuzzy nanoassemblies: toward layered polymeric multicomposites. Science 277(5330):1232–1237. doi:10.1126/science.277.5330.1232

Dong Y, Li X, Tian L, Bell T, Sammons R, Dong H (2011) Towards long-lasting antibacterial stainless steel surfaces by combining double glow plasma silvering with active screen plasma nitriding. Acta Biomater 7(1):447–457. doi:10.1016/j.actbio.2010.08.009

Dueland R, Spadaro JA, Rahn BA (1982) Silver antibacterial bone cement. Comparison with gentamicin in experimental osteomyelitis. Clin Orthop Relat Res 169:264–268

Elnathan R, Kwiat M, Patolsky F, Voelcker NH (2014) Engineering vertically aligned semiconductor nanowire arrays for applications in the life sciences. Nano Today 9(2):172–196. doi:10.1016/j.nantod.2014.04.001

Fadeeva E, Truong VK, Stiesch M, Chichkov BN, Crawford RJ, Wang J, Ivanova EP (2011) Bacterial retention on superhydrophobic titanium surfaces fabricated by femtosecond laser ablation. Langmuir 27(6):3012–3019. doi:10.1021/la104607g

Foster TJ, Höök M (1998) Surface protein adhesins of Staphylococcus aureus. Trends Microbiol 6(12):484–488. doi:10.1016/S0966-842X(98)01400-0

Foster TJ, Geoghegan JA, Ganesh VK, Höök M (2014) Adhesion, invasion and evasion: the many functions of the surface proteins of Staphylococcus aureus. Nat Rev Microbiol 12(1):49–62. doi:10.1038/nrmicro3161

Fox S, Wilkinson TS, Wheatley PS, Xiao B, Morris RE, Sutherland A, Simpson AJ, Barlow PG, Butler AR, Megson IL, Rossi AG (2010) NO-loaded Zn^{2+}-exchanged zeolite materials: a

potential bifunctional anti-bacterial strategy. Acta Biomater 6(4):1515–1521. doi:10.1016/j.actbio.2009.10.038
Fusetani N (2004) Biofouling and antifouling. Nat Prod Rep 21(1):94–104. doi:10.1039/b302231p
Fusetani N (2011) Antifouling marine natural products. Nat Prod Rep 28(2):400–410. doi:10.1039/c0np00034e
Gabriel M, Nazmi K, Veerman EC, Amerongen AVN, Zentner A (2006) Preparation of LL-37-grafted titanium surfaces with bactericidal activity. Bioconjug Chem 17(2):548–550. doi:10.1021/bc050091v
Gadenne V, Lebrun L, Jouenne T, Thebault P (2013) Antiadhesive activity of ulvan polysaccharides covalently immobilized onto titanium surface. Colloids Surf B: Biointerfaces 112(0):229–236. doi:10.1016/j.colsurfb.2013.07.061
Gerberich BG, Bhatia SK (2013) Tissue scaffold surface patterning for clinical applications. Biotechnol J 8(1):73–84. doi:10.1002/biot.201200131
Glinel K, Thebault P, Humblot V, Pradier C-M, Jouenne T (2012) Antibacterial surfaces developed from bio-inspired approaches. Acta Biomater 8(5):1670–1684. doi:10.1016/j.actbio.2012.01.011
Godoy-Gallardo M, Mas-Moruno C, Fernández-Calderón MC, Pérez-Giraldo C, Manero JM, Albericio F, Gil FJ, Rodríguez D (2014) Covalent immobilization of hLf1-11 peptide on a titanium surface reduces bacterial adhesion and biofilm formation. Acta Biomater 10(8):3522–3534. doi:10.1016/j.actbio.2014.03.026
Gong SQ, Niu LN, Kemp LK, Yiu CKY, Ryou H, Qi YP, Blizzard JD, Nikonov S, Brackett MG, Messer RLW, Wu CD, Mao J, Bryan Brister L, Rueggeberg FA, Arola DD, Pashley DH, Tay FR (2012) Quaternary ammonium silane-functionalized, methacrylate resin composition with antimicrobial activities and self-repair potential. Acta Biomater 8(9):3270–3282. doi:10.1016/j.actbio.2012.05.031
Gordon O, Slenters TV, Brunetto PS, Villaruz AE, Sturdevant DE, Otto M, Landmann R, Fromm KM (2010) Silver coordination polymers for prevention of implant infection: thiol interaction, impact on respiratory chain enzymes, and hydroxyl radical induction. Antimicrob Agents Chemother 54(10):4208–4218. doi:10.1128/AAC.01830-09
Grinthal A, Aizenberg J (2014) Mobile interfaces: liquids as a perfect structural material for multifunctional, antifouling surfaces. Chem Mater 26(1):698–708. doi:10.1021/cm402364d
Gristina AG (1987) Biomaterial-centered infection: microbial adhesion versus tissue integration. Science 237(4822):1588–1595. doi:10.1126/science.3629258
Gristina AG, Naylor P, Myrvik Q (1988) Infections from biomaterials and implants: a race for the surface. Med Prog Technol 14(3–4):205–224
Hajipour MJ, Fromm KM, Akbar Ashkarran A, Jimenez de Aberasturi D, Larramendi IRD, Rojo T, Serpooshan V, Parak WJ, Mahmoudi M (2012) Antibacterial properties of nanoparticles. Trends Biotechnol 30(10):499–511. doi:10.1016/j.tibtech.2012.06.004
Hammond PT (1999) Recent explorations in electrostatic multilayer thin film assembly. Curr Opin Colloid Interface Sci 4(6):430–442. doi:10.1016/S1359-0294(00)00022-4
Hammond PT (2004) Form and function in multilayer assembly: new applications at the nanoscale. Adv Mater 16(15):1271–1293. doi:10.1002/adma.200400760
Harris LG, Tosatti S, Wieland M, Textor M, Richards RG (2004) *Staphylococcus aureus* adhesion to titanium oxide surfaces coated with non-functionalized and peptide-functionalized poly(L-lysine)-grafted- poly(ethylene glycol) copolymers. Biomaterials 25(18):4135–4148. doi:10.1016/j.biomaterials.2003.11.033
Hasan J, Crawford RJ, Ivanova EP (2013) Antibacterial surfaces: the quest for a new generation of biomaterials. Trends Biotechnol 31(5):295–304. doi:10.1016/j.tibtech.2013.01.017
Hauck CR, Agerer F, Muenzner P, Schmitter T (2006) Cellular adhesion molecules as targets for bacterial infection. Eur J Cell Biol 85(3–4):235–242. doi:10.1016/j.ejcb.2005.08.002
Hebeish A, Abdelhady M, Youssef A (2013) TiO_2 nanowire and TiO_2 nanowire doped Ag-PVP nanocomposite for antimicrobial and self-cleaning cotton textile. Carbohydr Polym 91(2):549–559. doi:10.1016/j.carbpol.2012.08.068

Heidenau F, Mittelmeier W, Detsch R, Haenle M, Stenzel F, Ziegler G, Gollwitzer H (2005) A novel antibacterial titania coating: metal ion toxicity and in vitro surface colonization. J Mater Sci Mater Med 16(10):883–888. doi:10.1007/s10856-005-4422-3

Hempel F, Finke B, Zietz C, Bader R, Weltmann K-D, Polak M (2014) Antimicrobial surface modification of titanium substrates by means of plasma immersion ion implantation and deposition of copper. Surf Coat Technol. doi:10.1016/j.surfcoat.2014.01.027

Hoene A, Prinz C, Walschus U, Lucke S, Patrzyk M, Wilhelm L, Neumann HG, Schlosser M (2013) In vivo evaluation of copper release and acute local tissue reactions after implantation of copper-coated titanium implants in rats. Biomed Mater 8(3):035009. doi:10.1088/1748-6041/8/3/035009

Høiby N, Bjarnsholt T, Givskov M, Molin S, Ciofu O (2010) Antibiotic resistance of bacterial biofilms. Int J Antimicrob Agents 35(4):322–332. doi:10.1016/j.ijantimicag.2009.12.011

Holmberg KV, Abdolhosseini M, Li Y, Chen X, Gorr SU, Aparicio C (2013) Bio-inspired stable antimicrobial peptide coatings for dental applications. Acta Biomater 9(9):8224–8231. doi:10.1016/j.actbio.2013.06.017

Hu X, Neoh KG, Shi Z, Kang ET, Poh C, Wang W (2010) An in vitro assessment of titanium functionalized with polysaccharides conjugated with vascular endothelial growth factor for enhanced osseointegration and inhibition of bacterial adhesion. Biomaterials 31(34):8854–8863. doi:10.1016/j.biomaterials.2010.08.006

Huo K, Zhang X, Wang H, Zhao L, Liu X, Chu PK (2013) Osteogenic activity and antibacterial effects on titanium surfaces modified with Zn-incorporated nanotube arrays. Biomaterials 34(13):3467–3478. doi:10.1016/j.biomaterials.2013.01.071

Ip M, Lui SL, Poon VK, Lung I, Burd A (2006) Antimicrobial activities of silver dressings: an in vitro comparison. J Med Microbiol 55(1):59–63. doi:10.1099/jmm.0.46124-0

Ivanova EP, Hasan J, Truong VK, Wang JY, Raveggi M, Fluke C, Crawford RJ (2011) The influence of nanoscopically thin silver films on bacterial viability and attachment. Appl Microbiol Biotechnol 91(4):1149–1157. doi:10.1007/s00253-011-3195-5

Ivanova EP, Hasan J, Webb HK, Truong VK, Watson GS, Watson JA, Baulin VA, Pogodin S, Wang JY, Tobin MJ, Löbbe C, Crawford RJ (2012) Natural bactericidal surfaces: mechanical rupture of *Pseudomonas aeruginosa* cells by cicada wings. Small 8(16):2489–2494. doi:10.1002/smll.201200528

Ivanova EP, Hasan J, Webb HK, Gervinskas G, Juodkazis S, Truong VK, Wu AHF, Lamb RN, Baulin VA, Watson GS, Watson JA, Mainwaring DE, Crawford RJ (2013) Bactericidal activity of black silicon. Nat Commun 4:2838. doi:10.1038/ncomms3838

Jahed Z, Molladavoodi S, Seo BB, Gorbet M, Tsui TY, Mofrad MRK (2014) Cell responses to metallic nanostructure arrays with complex geometries. Biomaterials 35(34):9363–9371. doi:10.1016/j.biomaterials.2014.07.022

Jeon H, Simon CG, Kim G (2014) A mini-review: cell response to microscale, nanoscale, and hierarchical patterning of surface structure. J Biomed Mater Res B Appl Biomater 102(7):1580–1594

Joo Y-K, Zhang S-H, Yoon J-H, Cho T-Y (2009) Optimization of the adhesion strength of arc ion plating TiAlN films by the Taguchi method. Materials 2(2):699–709. doi:10.3390/ma2020699

Kang B-M, Lim Y-S (2014) Antibacterial properties of TiAgN and ZrAgN thin film coated by physical vapor deposition for medical applications. Trans Electr Electron Mater 15(5):275–278. doi:10.4313/TEEM.2014.15.5.275

Kawashita M, Tsuneyama S, Miyaji F, Kokubo T, Kozuka H, Yamamoto K (2000) Antibacterial silver-containing silica glass prepared by sol-gel method. Biomaterials 21(4):393–398. doi:10.1016/S0142-9612(99)00201-X

Kim W, Ng JK, Kunitake ME, Conklin BR, Yang P (2007) Interfacing silicon nanowires with mammalian cells. J Am Chem Soc 129(23):7228–7229. doi:10.1021/ja071456k

Lambris JD, Ricklin D, Geisbrecht BV (2008) Complement evasion by human pathogens. Nat Rev Microbiol 6(2):132–142. doi:10.1038/nrmicro1824

Lavernia C, Lee DJ, Hernandez VH (2006) The increasing financial burden of knee revision surgery in the United States. Clin Orthop Relat Res 446:221–226. doi:10.1097/01.blo.0000214424.67453.9a

Lemire JA, Harrison JJ, Turner RJ (2013) Antimicrobial activity of metals: mechanisms, molecular targets and applications. Nat Rev Microbiol 11(6):371–384. doi:10.1038/nrmicro3028

Li H, Cui Q, Feng B, Wang J, Lu X, Weng J (2013) Antibacterial activity of TiO 2 nanotubes: influence of crystal phase, morphology and Ag deposition. Appl Surf Sci 284:179–183

Li J, Liu X, Qiao Y, Zhu H, Ding C (2014) Antimicrobial activity and cytocompatibility of Ag plasma-modified hierarchical TiO_2 film on titanium surface. Colloids Surf B: Biointerfaces 113:134–145. doi:10.1016/j.colsurfb.2013.08.030

Linford MR, Auch M, Möhwald H (1998) Nonmonotonic effect of ionic strength on surface dye extraction during dye-polyelectrolyte multilayer formation. J Am Chem Soc 120(1):178–182. doi:10.1021/ja972133z

Lu T, Qiao Y, Liu X (2012) Surface modification of biomaterials using plasma immersion ion implantation and deposition. Interface Focus 2(3):325–336. doi:10.1098/rsfs.2012.0003

Lv Q, Feng Q (2006) Preparation of 3-D regenerated fibroin scaffolds with freeze drying method and freeze drying/foaming technique. J Mater Sci Mater Med 17(12):1349–1356

Lv W, Luo J, Deng Y, Sun Y (2013) Biomaterials immobilized with chitosan for rechargeable antimicrobial drug delivery. J Biomed Mater Res A 101 A(2):447–455

Maddikeri RR, Tosatti S, Schuler M, Chessari S, Textor M, Richards RG, Harris LG (2008) Reduced medical infection related bacterial strains adhesion on bioactive RGD modified titanium surfaces: a first step toward cell selective surfaces. J Biomed Mater Res A 84(2):425–435. doi:10.1002/jbm.a.31323

Maury F, Mungkalasiri J, Bedel L, Emieux F, Dore J, Renaud FN (2014) Comparative study of antibacterial efficiency of M-TiO_2 (M Ag, Cu) thin films grown by CVD. Key Eng Mater 617:127–130. doi:10.4028/www.scientific.net/KEM.617.127

McCloskey AP, Gilmore BF, Laverty G (2014) Evolution of antimicrobial peptides to self-assembled peptides for biomaterial applications. Pathogens 3(4):792–821. doi:10.3390/pathogens3040791

McLean RJC, Hussain AA, Sayer M, Vincent PJ, Hughes DJ, Smith TJN (1993) Antibacterial activity of multilayer silver-copper surface films on catheter material. Can J Microbiol 39(9):895–899

Mei L, Ren Y, Loontjens TJA, van der Mei HC, Busscher HJ (2012) Contact-killing of adhering streptococci by a quaternary ammonium compound incorporated in an acrylic resin. Int J Artif Organs 35(10):854–863. doi:10.5301/ijao.5000149

Meng J, Zhang P, Wang S (2014) Recent progress in biointerfaces with controlled bacterial adhesion by using chemical and physical methods. Chem Asian J 9(8):2004–2016. doi:10.1002/asia.201402200

Minagar S, Berndt CC, Wang J, Ivanova E, Wen C (2012) A review of the application of anodization for the fabrication of nanotubes on metal implant surfaces. Acta Biomater 8(8):2875–2888. doi:10.1016/j.actbio.2012.04.005

Montanaro L, Speziale P, Campoccia D, Ravaioli S, Cangini I, Pietrocola G, Giannini S, Arciola CR (2011) Scenery of Staphylococcus implant infections in orthopedics. Future Microbiol 6(11):1329–1349. doi:10.2217/fmb.11.117

Nablo BJ, Prichard HL, Butler RD, Klitzman B, Schoenfisch MH (2005) Inhibition of implant-associated infections via nitric oxide release. Biomaterials 26(34):6984–6990. doi:10.1016/j.biomaterials.2005.05.017

Neoh KG, Hu X, Zheng D, Kang ET (2012) Balancing osteoblast functions and bacterial adhesion on functionalized titanium surfaces. Biomaterials 33(10):2813–2822. doi:10.1016/j.biomaterials.2012.01.018

Nomiya K, Tsuda K, Sudoh T, Oda M (1997) Ag(I)-N bond-containing compound showing wide spectra in effective antimicrobial activities: polymeric silver(I) imidazolate. J Inorg Biochem 68(1):39–44. doi:10.1016/S0162-0134(97)00006-8

Ogaki R, Alexander M, Kingshott P (2010) Chemical patterning in biointerface science. Mater Today 13(4):22–35. doi:10.1016/S1369-7021(10)70057-2

Olson ME, Ceri H, Morck DW, Buret AG, Read RR (2002) Biofilm bacteria: formation and comparative susceptibility to antibiotics. Can J Vet Res 66(2):86–92

Paasche G, Ceschi P, Löbler M, Rösl C, Gomes P, Hahn A, Rohm HW, Sternberg K, Lenarz T, Schmitz KP, Barcikowski S, Stöver T (2011) Effects of metal ions on fibroblasts and spiral ganglion cells. J Neurosci Res 89(4):611–617. doi:10.1002/jnr.22569

Pant HR, Pant B, Sharma RK, Amarjargal A, Kim HJ, Park CH, Tijing LD, Kim CS (2013) Antibacterial and photocatalytic properties of Ag/TiO_2/ZnO nano-flowers prepared by facile one-pot hydrothermal process. Ceram Int 39(2):1503–1510. doi:10.1016/j.ceramint.2012.07.097

Patti JM, Allen BL, McGavin MJ, Höök M (1994) MSCRAMM-mediated adherence of microorganisms to host tissues. Annu Rev Microbiol 48:585–617

Percival S, Bowler P, Russell D (2005) Bacterial resistance to silver in wound care. J Hosp Infect 60(1):1–7. doi:10.1016/j.jhin.2004.11.014

Pegalajar-Jurado A, Easton CD, Styan KE, McArthur SL (2014) Antibacterial activity studies of plasma polymerised cineole films. J Mater Chem B 2(31):4993–5002

Pourbaix M (1984) Electrochemical corrosion of metallic biomaterials. Biomaterials 5(3):122–134. doi:10.1016/0142-9612(84)90046-2

Prantl L, Bürgers R, Schreml S, Zellner J, Gosau M (2010) A novel antibacterial silicone implant material with short- and long-term release of copper ions. Plast Reconstr Surg 125(2):78e–80e. doi:10.1097/PRS.0b013e3181c2a708

Rautray TR, Narayanan R, Kwon TY, Kim KH (2010) Surface modification of titanium and titanium alloys by ion implantation. J Biomed Mater Res B Appl Biomater 93(2):581–591. doi:10.1002/jbm.b.31596

Richards RME (1981) Antimicrobial action of silver nitrate. Microbios 31(124):83–91

Rivero PJ, Urrutia A, Goicoechea J, Zamarreño CR, Arregui FJ, Matías IR (2011) An antibacterial coating based on a polymer/sol-gel hybrid matrix loaded with silver nanoparticles. Nanoscale Res Lett 6(1):1–7. doi:10.1186/1556-276X-6-305

Robinson JT, Jorgolli M, Shalek AK, Yoon MH, Gertner RS, Park H (2012) Vertical nanowire electrode arrays as a scalable platform for intracellular interfacing to neuronal circuits. Nat Nanotechnol 7(3):180–184. doi:10.1038/nnano.2011.249

Ruggieri MR, Hanno PM, Levin RM (1987) Reduction of bacterial adherence to catheter surface with heparin. J Urol 138(2):423–426

Salwiczek M, Qu Y, Gardiner J, Strugnell RA, Lithgow T, McLean KM, Thissen H (2014) Emerging rules for effective antimicrobial coatings. Trends Biotechnol 32(2):82–90. doi:10.1016/j.tibtech.2013.09.008

Samanovic MI, Ding C, Thiele DJ, Darwin KH (2012) Copper in microbial pathogenesis: meddling with the metal. Cell Host Microbe 11(2):106–115. doi:10.1016/j.chom.2012.01.009

Schaer TP, Stewart S, Hsu BB, Klibanov AM (2012) Hydrophobic polycationic coatings that inhibit biofilms and support bone healing during infection. Biomaterials 33(5):1245–1254. doi:10.1016/j.biomaterials.2011.10.038

Shi Z, Neoh KG, Kang ET, Poh C, Wang W (2008) Bacterial adhesion and osteoblast function on titanium with surface-grafted chitosan and immobilized RGD peptide. J Biomed Mater Res A 86(4):865–872. doi:10.1002/jbm.a.31648

Shirai T, Tsuchiya H, Shimizu T, Ohtani K, Zen Y, Tomita K (2009) Prevention of pin tract infection with titanium-copper alloys. J Biomed Mater Res B Appl Biomater 91(1):373–380. doi:10.1002/jbm.b.31412

Siedenbiedel F, Tiller JC (2012) Antimicrobial polymers in solution and on surfaces: overview and functional principles. Polymer 4(1):46–71. doi:10.3390/polym4010046

Song Z, Borgwardt L, Høiby N, Wu H, Sørensen TS, Borgwardt A (2013) Prosthesis infections after orthopedic joint replacement: the possible role of bacterial biofilms. Orthop Rev 5:e14. doi:10.4081/or.2013.e14

Stafford SL, Bokil NJ, Achard ME, Kapetanovic R, Schembri MA, McEwan AG, Sweet MJ (2013) Metal ions in macrophage antimicrobial pathways: emerging roles for zinc and copper. Biosci Rep 33(4):e00049

Subbiahdoss G, Pidhatika B, Coullerez G, Charnley M, Kuijer R, van der Mei HC, Textor M, Busscher HJ (2010) Bacterial biofilm formation versus mammalian cell growth on titanium-based mono-and bi-functional coatings. Eur Cell Mater 19:205–213

Talebian N, Doudi M, Kheiri M (2014) The anti-adherence and bactericidal activity of sol–gel derived nickel oxide nanostructure films: solvent effect. J Sol-Gel Sci Technol 69(1):172–182. doi:10.1007/s10971-013-3201-8

Trivedi P, Gupta P, Srivastava S, Jayaganthan R, Chandra R, Roy P (2014) Characterization and in vitro biocompatibility study of Ti–Si–N nanocomposite coatings developed by using physical vapor deposition. Appl Surf Sci 293(0):143–150. doi:10.1016/j.apsusc.2013.12.119

Truong VK, Webb HK, Fadeeva E, Chichkov BN, Wu AHF, Lamb R, Wang JY, Crawford RJ, Ivanova EP (2012) Air-directed attachment of coccoid bacteria to the surface of superhydrophobic lotus-like titanium. Biofouling 28(6):539–550. doi:10.1080/08927014.2012.694426

Varghese S, Elfakhri S, Sheel D, Sheel P, Bolton F, Foster H (2013) Novel antibacterial silver-silica surface coatings prepared by chemical vapour deposition for infection control. J Appl Microbiol 115(5):1107–1116. doi:10.1111/jam.12308

Vasilev K, Cook J, Griesser HJ (2009) Antibacterial surfaces for biomedical devices. Expert Rev Med Devices 6(5):553–567. doi:10.1586/erd.09.36

Visai L, De Nardo L, Punta C, Melone L, Cigada A, Imbriani M, Arciola CR (2011) Titanium oxide antibacterial surfaces in biomedical devices. Int J Artif Organs 34(9):929–946. doi:10.5301/ijao.5000050

Wan YZ, Raman S, He F, Huang Y (2007a) Surface modification of medical metals by ion implantation of silver and copper. Vacuum 81(9):1114–1118. doi:10.1016/j.vacuum.2006.12.011

Wan YZ, Xiong GY, Liang H, Raman S, He F, Huang Y (2007b) Modification of medical metals by ion implantation of copper. Appl Surf Sci 253(24):9426–9429. doi:10.1016/j.apsusc.2007.06.031

Webb HK, Hasan J, Truong VK, Crawford RJ, Ivanova EP (2011) Nature inspired structured surfaces for biomedical applications. Curr Med Chem 18(22):3367–3375. doi:10.2174/092986711796504673

Webb HK, Crawford RJ, Ivanova EP (2014) Wettability of natural superhydrophobic surfaces. Adv Colloid Interf Sci 210:58–64. doi:10.1016/j.cis.2014.01.020

Whitehouse JD, Deborah Friedman N, Kirkland KB, Richardson WJ, Sexton DJ (2002) The impact of surgical-site infections following orthopedic surgery at a community hospital and a university hospital: adverse quality of life, excess length of stay, and extra cost. Infect Control Hosp Epidemiol 23(4):183–189. doi:10.1086/502033

Wilkinson M, Kafizas A, Bawaked SM, Obaid AY, Al-Thabaiti SA, Basahel SN, Carmalt CJ, Parkin IP (2013) Combinatorial atmospheric pressure chemical vapor deposition of graded TiO2–VO2 mixed-phase composites and their dual functional property as self-cleaning and photochromic window coatings. ACS Comb Sci 15(6):309–319

Williams JF, Worley SD (2000) Infection-resistant nonleachable materials for urologic devices. J Endourol 14(5):395–400. doi:10.1089/end.2000.14.395

Wong SY, Li Q, Veselinovic J, Kim B-S, Klibanov AM, Hammond PT (2010) Bactericidal and virucidal ultrathin films assembled layer by layer from polycationic N-alkylated polyethylenimines and polyanions. Biomaterials 31(14):4079–4087. doi:10.1016/j.biomaterials.2010.01.119

Wong CL, Tan YN, Mohamed AR (2011) A review on the formation of titania nanotube photocatalysts by hydrothermal treatment. J Environ Manag 92(7):1669–1680

Yildirimer L, Thanh NTK, Loizidou M, Seifalian AM (2011) Toxicological considerations of clinically applicable nanoparticles. Nano Today 6(6):585–607. doi:10.1016/j.nantod.2011.10.001

Yoshinari M, Oda Y, Kato T, Okuda K (2001) Influence of surface modifications to titanium on antibacterial activity in vitro. Biomaterials 22(14):2043–2048. doi:10.1016/S0142-9612(00)00392-6

Yuan S, Wan D, Liang B, Pehkonen S, Ting Y, Neoh K, Kang E (2011) Lysozyme-coupled poly (poly (ethylene glycol) methacrylate)– stainless steel hybrids and their antifouling and antibacterial surfaces. Langmuir 27(6):2761–2774. doi:10.1021/la104442f

Yue C, Kuijer R, Kaper HJ, van der Mei HC, Busscher HJ (2014) Simultaneous interaction of bacteria and tissue cells with photocatalytically activated, anodized titanium surfaces. Biomaterials 35(9):2580–2587

Zaborowska M, Welch K, Brånemark R, Khalilpour P, Engqvist H, Thomsen P, Trobos M (2014) Bacteria-material surface interactions: methodological development for the assessment of implant surface induced antibacterial effects. J Biomed Mater Res B Appl Biomater. doi:10.1002/jbm.b.33179

Zhang W, Li Y, Niu J, Chen Y (2013) Photogeneration of reactive oxygen species on uncoated silver, gold, nickel, and silicon nanoparticles and their antibacterial effects. Langmuir 29(15):4647–4651. doi:10.1021/la400500t

Zhao L, Chu PK, Zhang Y, Wu Z (2009) Antibacterial coatings on titanium implants. J Biomed Mater Res Part B Appl Biomater 91(1):470–480. doi:10.1002/jbm.b.31463

Zhao B, Van Der Mei HC, Subbiahdoss G, De Vries J, Rustema-Abbing M, Kuijer R, Busscher HJ, Ren Y (2014) Soft tissue integration versus early biofilm formation on different dental implant materials. Dent Mater 30(7):716–727. doi:10.1016/j.dental.2014.04.001

Zobrist C, Sobocinski J, Lyskawa J, Fournier D, Miri V, Traisnel M, Jimenez M, Woisel P (2011) Functionalization of titanium surfaces with polymer brushes prepared from a biomimetic RAFT agent. Macromolecules 44(15):5883–5892

Chapter 7
Cytotoxic Effects and Biocompatibility of Antimicrobial Materials

Olha Bazaka and Kateryna Bazaka

Abstract The rising demand for medical implants for ageing populations and ongoing advancements in medical technology continue to drive the use of implantable devices. Higher implant usage has a consequent increased incidence of implant-related infections, and associated prolonged patient care, pain and loss of limb and other organ function. Numerous antibacterial surfaces have been designed that prevent the onset of biofilm formation, thus reducing or preventing implant-associated infections through inhibiting bacterial adhesion or by killing the organisms that successfully attach to the surface of the implant. Other surfaces have been designed to stimulate a local immune response, promoting the natural clearing of the invading pathogen. The desired antibacterial effects are typically achieved by modulating the surface chemistry and morphology of the implant material, by means of the controlled release of pharmacological agents and bioactive compounds from the surface of the material, or by a combination of both processes. An important issue for any type of antibacterial surface modification lies in balancing the non-fouling, bacteriostatic or bactericidal effects against local and systemic biocompatibility. In this chapter, we will first describe the concept of biocompatibility and its evolution, from devices that do not evoke a negative host response to those that actively drive host regeneration. We will then review the challenges associated with merging the need for an implant material to withstand a bacterial load with those associated with supporting function restoration and tissue healing.

Keywords Bacteriostatic • Bactericidal • Systemic biocompatibility • Cytotoxicity • Inflammation • Host regeneration • Tissue healing

O. Bazaka (✉) • K. Bazaka
College of Science, Technology and Engineering, James Cook University, Townsville, QLD 4810, Australia

E.P. Ivanova, R.J. Crawford (eds.), *Antibacterial Surfaces*,
DOI 10.1007/978-3-319-18594-1_7

7.1 Introduction

Any surgical intervention carries a risk of infection. Whenever the skin, a natural barrier to infection, is perforated or incised, a significant opportunity is created for infective agents to enter the body. The likelihood of developing an infection is dependent on a multitude of procedure- and patient-related factors (Phillips et al. 2014). These include the type, invasiveness, duration and complexity of the surgery (Olsen et al. 2008; Neumayer et al. 2007). Minimally invasive surgery techniques have been found to notably reduce post-operative wound infections by as much as tenfold in the case of minimally invasive versus open spinal surgery (O'Toole et al. 2009).

The skill of the operating team, the cleanness of the surgical environment and the use of pre- and post-surgery antibiotics have also shown to vastly reduce, if not fully eliminate the patient's risk of being infected (Ridgeway et al. 2005; Sharaf et al. 2011). Indeed, it is very difficult to create a perfectly bacteria-free operating theatre and thus avoid at least low-cell-load bacterial contamination of the wound space. Whilst strategies can be used to lower the incidence of surgical wound infections that may occur though one process, this same strategy may inadvertently increase the risk of infection through another pathway. For example, intraoperative warming has been shown to reduce postoperative infections in patients undergoing major orthopedic, general, or urologic surgery by maintaining a cutaneous blood flow, increasing tissue viability and reducing the incidence of pressure sores. When intra-operative warming has been performed using forced-air devices, however, the moving air may act as a vector for the transfer of infectious agents, causing unwanted airflow disturbances (Weissman and Murray 2013; Augustine 2014).

Interestingly, in a large proportion of surgical site infections, the pathogens responsible are not exogenous, but emerge from the normal endogenous flora of the patient. *Staphylococcus aureus*, coagulase-negative staphylococci, *Pseudomonas aeruginosa*, Enterococcus spp. and *Escherichia coli* are among most commonly implicated pathogens (Owens and Stoessel 2008). The act of surgery facilitates the transport of these microorganisms from their habitat, e.g. skin or gastrointestinal tract, to the sterile sites of the body, such as spinal cord or muscle tissue (Thakkar et al. 2014). An infection elsewhere in the body, whether being a pre-existing condition or having been acquired post-surgery, can also serve as an internal source of infection (Claridge et al. 2014).

An individual's risk of developing an infection depends greatly on the patient's state of health, and particularly their ability to resist infection. Immunocompromised patients, such as the elderly, pregnant women, those suffering from chronic illnesses, those undergoing cancer treatment, recovering from a recent illness or invasive intervention, or having certain genetic disorders, may succumb to significantly lower bacterial loads or develop infections caused by opportunistic bacteria (Neumayer et al. 2007; Ridgeway et al. 2005). For example, patients suffering from hypoalbuminemia prior to gastrointestinal surgery are more likely to develop deeper surgical site infections and recover more slowly compared to patients whose pre-operative albumin levels were higher (Hennessey et al. 2010). According to a review study published by Pittet et al., patients being treated for breast cancer and undergoing a mastectomy experienced an increased risk of surgical site infection from an average rate of 2.5 % up to 53 %. This increase can be attributed to postoperative tissue ischemia and delayed wound healing (Pittet et al. 2005).

7.1.1 *Biomaterial-Associated Infections and Inflammation*

The introduction of an implant into the body further predisposes the host to infection and inflammation. For one, tissue that has already been injured by the surgical intervention is subjected to further physical damage from implant insertion and positioning. Subsequent to the insertion, the implant may continue to inflict physical damage to the adjacent tissues, putting excessive pressure on the surrounding tissues, rubbing against the tissues or by breaching epithelial or mucosal barriers. This breach may provide a means for bacteria to enter host tissues, whereas the subsequent physical trauma may impair the ability of the tissue to recover and to resist bacterial invasion (Chandorkar et al. 2015).

A persistent sterile inflammatory response, in itself, presents a significant issue for all medical implants, irrespective of their type and method of implantation. For example, a chronic foreign body response to devices implanted into brain tissue affects the electrophysiological and neurobiological activity of nearby neurons and the neural circuitry by a variety of neuro-inflammatory mechanisms, including disruption of the blood brain barrier, demyelination and reduction in density of local nerve fibers and neuronal cell bodies, and changes in the local ionic milieu (Skousen et al. 2015). Continuing physical damage and chronic infection can lead to tissue necrosis and permanent changes in the tissue surrounding the implant.

The surface of the implant may serve as a vector for the introduction of the microorganisms into the body. Although in developed countries, this issue is rarely observed due to highly controlled sterilization practices, the sterilization compliance remains an issue in countries with limited-resources (Rosenthal 2008). It is important to note that implant sterilization does not automatically render it free from agents that can induce a pro-inflammatory response, namely the endotoxins and micro- and nano-sized particles. Endotoxins are the lipopolysaccharides in the cell membrane of Gram-negative bacteria (Vetten et al. 2014). Residual endotoxin contaminations can vary significantly, depending on the method used for sterilizing the biomaterial.

For example, microelectrodes sterilized by autoclave, dry heat, or ethylene oxide gas, retained residual endotoxins of 0.55 EU per mL, 0.22 EU per mL, and 0.11 EU per mL, respectively (Ravikumar et al. 2014). At early stages of implantation into neural tissue, the extent of microglia/macrophage activation is directly related to the level of endotoxins present on the implant surface, whereas the degree of astrogliosis, neuronal loss, and blood-brain barrier dysfunction exhibit a threshold-dependent response. The effect of endotoxin levels diminishes with longer implantation periods, suggesting that the endotoxin contributes to the initial, but not chronic neuro-inflammation.

Just as initially sterile implant surfaces can induce a foreign body response, the abiotic surface of the implant provides excellent grounds for secondary colonization of the implant surface via the hematogenous dissemination from a remote focus of the infection. Indeed, conventional (non-antibacterial) biomaterials are not able to prevent bacteria from permanently attaching to their surfaces and forming a biofilm, nor activating a host immune response. Once colonized, the implant surface can not only lead to peri-implant infection but also serve as a reservoir from which bacteria can spread to other parts of the body (Claridge et al. 2014).

Implant-associated infections are difficult to treat with antibiotics alone and often necessitate the removal of the implant. The protective nature of a biofilm significantly reduces the efficacy of both pharmacological therapies and host defense mechanisms (Lisanti et al. 2015; Khan et al. 2014). In addition, as most biomaterials do not react to being colonized, the infection is often detected only after the biofilm has been formed and the pathogen activity begins to impact on the surrounding tissues. Although some progress has been made in the development of strategies for dissolution of biofilms that have already been established, targeting of the biofilm matrix for degradation, interference with the biofilm regulation (Hogan et al. 2015) and removal of biofilm-residing bacteria remains a challenge.

7.1.2 Resistance and Persistence of Pathogenic Bacteria

When resident in a biofilm, certain species of bacteria, such as staphylococci, exhibit increased mutation ability and enhanced ability to acquire/disseminate plasmid-borne antibiotic resistance determinants by horizontal gene transfer (Savage et al. 2013; Ryder et al. 2012). Compared to planktonic cultures, the mutation ability of *S. aureus* and *S. epidermidis* biofilm cultures increased up to 60-fold and 4-fold, respectively. Such enhanced spontaneous mutation accelerates the emergence of heritable antibiotic resistance and horizontal gene transfer ensures that this resistance spreads throughout the biofilm community. Mixed-species biofilms can also be more resilient to the action of antibiotics due to phenomenon known as indirect pathogenicity, where an antibiotic-resistant microorganism of low intrinsic virulence shields an antibiotic-sensitive pathogen from eradication (O'Connell et al. 2006). For example, antibiotics such as penicillins and cephalosporins can be inactivated via hydrolysis of the cyclic amide bond by beta-lactamase-producing bacteria, thereby preventing these antimicrobials from acting upon susceptible microorganisms. With an increasing prevalence of multidrug resistant bacteria, identifying antibiotic agents that would be able to overcome multidrug resistance in biofilm-dwelling bacteria is very challenging.

In addition to enzymes, antibiotic-sensitive strains can also benefit from the action of multifunctional extracellular polymers produced by other bacteria within the community. For example, *P. aeruginosa* produces Psl polysaccharide, which acts as molecular glue to enable bacterial attachment, self-organization of the attached cells into micro-colonies, and communal biofilm organization (Zhao et al. 2013a). As such, Psl provides a generic first line of defense against antibiotics with diverse biochemical properties during the initial stages of biofilm development. In mixed-species biofilms, Psl mediated matrix protection is extended to non-Psl-producing antibiotic-sensitive *E. coli* and *S. aureus* cells (Billings et al. 2013). Reducing surface adhesion and the production of bacterial adhesins are often a strategy of choice to prevent biofilm formation and implant-related infections.

Even in a population of genetically identical antibiotic-sensitive bacterial cells, not all cells will die as a result of an antibiotic treatment, even at a sufficiently high antimicrobial dose; the killing efficacy of penicillin depends on the physiological state of the cell, with rapidly growing bacteria being killed most efficiently, followed by slowly growing cells, whereas dormant (non-growing) bacteria are generally not affected by the drug (Tuomanen et al. 1986; Gerdes and Maisonneuve 2012). Even in an actively growing population, however, not all cells will succumb to penicillin treatment. While most of the population will be killed, a fraction of this genetically homogeneous microbial population may persist. In contrast to resistant mutants, persistent microorganisms do not acquire resistance to the said antibiotic (Balaban et al. 2004), which confirms that persistence is a phenotypic and not genotypic effect.

Persistence stems from inherent phenotypic heterogeneity within a microbial population, where distinct subpopulations with different growth rates co-exist: normal cells and persister (slow-growing) cells. Normal cells ensure population survival under normal conditions, whereas persisters provide fitness benefit under stressed conditions, e.g. antibiotic load (Patra and Klumpp 2013). The phenotype switching is reversible, with the population regrown from the persister cells possessing both normal and persister cells. Peptidoglycan plasticity is a key factor in drug sensing and signaling to mitigate antibiotic exposure, where peptidoglycan remodeling enzymes are intimately involved in dormancy and resuscitation phenomena (Cava and de Pedro 2014; Amoroso et al. 2012; Courvalin 2006). In a dormant organism, the presence of specific peptidoglycan fragments can trigger the resuscitation pathway directly; peptidoglycan can also function as a substrate for the endogenous resuscitation-promoting factors, whereby it is hydrolyzed to liberate stimulatory fragments of different size (Nikitushkin et al. 2013). In addition to diversifying the growth rate to ensure population survival, bacteria can gain survival advantages from differentiation into different shapes (i.e. from bacillary to coccoid morphology) and into a filamentous morphology (Justice et al. 2014; Friedlander et al. 2013).

At present, the molecular mechanisms underlying bacterial persistence remain undiscovered (Maisonneuve et al. 2011), however it is believed that gene pairs encoding cognate toxin-antitoxins present in virtually all bacteria control reversible bacterial growth arrest (Holden 2015; Germain et al.). Equally, little is understood about the relationship between the recurring use of antibiotics, persistence and development of antibiotic resistance. In the future, the persistence-enabling toxin-antitoxin and associated signaling mechanisms may be targeted to either prevent persisters from arising or to transition them into a normal and thus antibiotic-susceptible state (Holden 2015).

7.1.3 Preventing Implant-Associated Infections via Implant Modification

It is evident that antibiotic prophylaxis and tightly controlled operating environment does not prevent an implant-associated infection from occurring. Furthermore, once established on the surface of the implant, a biofilm is difficult to eradicate with systemic antibiotics alone. The efficacy of these antibiotics is also diminishing, with antibiotic resistance spreading faster than the rate at which new antimicrobial compounds, such as teixobactin, can be discovered, evaluated and clinically introduced (Ling et al. 2015). Besides, persistent cells may remain on the surface of the implant, giving rise to chronic infections, and therefore the most logical step would be to localize the antibacterial activity to the surface of the implantable material itself.

Over the years, numerous strategies have been devised to prevent implant-associated infections by targeting different stages of biofilm formation and molecular and cellular mechanisms of bacterial pathogenesis (Grainger et al. 2013). These include various specific and non-specific anti-adhesive surfaces, and controlled systemic or localized drug delivery to interrupt cell-cell communication, inhibit bacterial aggregation and biofilm formation, or eliminate bacteria directly.

There are several important aspects that need to be considered when imparting antibacterial properties, whether they are physical or chemical features, onto the surface of the implant. An obvious factor is that the bacteria-repelling, bacteriostatic or bactericidal property should be maintained for the intended period of time under *in vivo* conditions, considering that *in vivo* environment may vary significantly over time and between patients. For example, if the anti-infective effect is due to the specific surface topographical features of the implants, these features should remain available and not be masked by protein fouling or cell debris.

The selected processing methodology should be compatible with the type of the implant material/s and not undermine the physico-chemical and mechanical integrity of the modified device. Neither should this modification destabilize the interactions between the individual components and/or different materials that may constitute a device or be a part of the fixative system, e.g. the adhesion of cement to the metallic component of the bone implant should be maintained. The modification itself should be mechanically robust to withstand normal pre-, intra-, and post-operative handling and operation. For example, an antibacterial coating on a load-bearing implant will experience a wide range of mechanical stresses and strains.

Critically, the modified implant should remain biocompatible and retain its ability to perform its intended function with an appropriate host response. Host response is triggered by the trauma of the surgical and implantation process, and then maintained by the physical, chemical, biological and mechanical interactions between the resident implant and the surrounding tissues. Depending on the nature of the implant and the stage of implantation, host response will include inflammatory and wound healing responses, foreign body reactions, or fibrous encapsulation (Anderson 2001).

7.1.4 Host Response and Implant Biocompatibility

Initiated by the action of implantation of a foreign body into a host, the tissue response continuum comprises a diverse ensemble of biological responses, including early events, such as injury, blood-implant interactions, provisional matrix formation, and acute inflammation, and later events, e.g. chronic inflammation, tissue granulation, foreign body reaction and fibrosis (Anderson 2001; Karlsson et al. 2014).

The extent of implantation injury is associated with the placement of the implant, its dimensions, weight and configuration, as well as its mechanical (e.g. flexible vs rigid) and surface (e.g. low-friction vs. abrasive) properties. Most implant-related anti-infective strategies are confined to the surface of the material, potentially altering the surface properties and three-dimensional configuration of the implant.

The injury incites a wound healing cascade, beginning with initial platelet adhesion to the injury site, followed by platelet activation and aggregation. In turn, this initiates a coagulation cascade, leading to clot formation as a measure to manage blood loss and minimise the risk of microbial infection. Clot formation is followed by highly mobile neutrophil granulocytes infiltrating the injury site as part of innate immune responses. Phagocytosis ensues clearing the debris of damaged tissues and pathogens that may have been introduced into the peri-implant space. Concurrently, phagocytic cells release growth factors to stimulate tissue regeneration, promoting the formation of granulation tissue.

Where an injury results in only exudative inflammation, inflammation resolution and restoration of the original tissue structure proceeds relatively quickly. On the other hand, dominance of pro-inflammatory over generative-type macrophages at the injury sites with significant tissue necrosis and/or loss of basement membrane structures may lead granulation tissue to grow into the inflammatory exudate. As a result, instead of restoration of original tissue structure, organisation and development of fibrous tissue takes place. It is important that even in the case of tissue restoration, re-grown cells may differ in their growth and differentiation behaviour from the original (pre-surgery) tissue. Regenerated cells can differ in their type, size, number, and function level; their ability to produce intra- an extra-cellular polymers can also change.

In addition to the nature of the injury (e.g. skill of the surgeon) and the health status of the host (e.g. suppressed proliferative capacity of cells), the nature of the implant may significantly affect the hemostasis, inflammation, repair, and remodeling processes, and thus directly influence the tissue healing outcomes. For example, copper-based coatings have been shown to effectively reduce infections on the surfaces of titanium implants. Implantation of copper-modified titanium implants has been associated with increased total and tissue macrophages and MHC-class-II-positive cells in a murine model (Andreas et al. 2013). As the tissue reactions persisted beyond the Cu release, it was concluded that surface-bound Cu may have also contributed to the increased local inflammatory response. In Cu/Ti coatings, an increasing Cu fraction led to an increase in the degree of hemolysis, limited break down of platelets and reduction in platelet adhesion (Liu et al. 2012a). At Cu concentrations above 10 wt.%, significant inhibition of endothelial cell adhesion and proliferation was observed. Given the important role of chemical factors released by plasma and cells in mediating the inflammatory response, implant-released chemicals may interfere with correct tissue regeneration (Yu et al. 2013). Furthermore, the lysosomal proteases and reactive oxygen species abundant at the site of inflammation may promote degradation of the implant material, thus further exacerbating the release of the active compounds.

The importance of the provisional matrix formation on tissue integration should not be underestimated (Anitua et al. 2014; Abrahamsson et al. 2004). The matrix is composed of many components; fibronectin, collagen and thrombospondin bind to fibrin and the platelet granule components of the initial clot. A major constituent of the matrix, fibrin induces neovascularization, promotes cell attachment (Thomson et al. 2013), and provides a scaffold for cell migration and tissue formation. The matrix also releases a wide array of mitogens, chemoattractants, cytokines, and growth factors that regulates cellular proliferation. The formation of the provisional matrix may be considerably altered by the surface properties of the materials (Tejero et al. 2014; Lang et al. 2011), and thus needs to be considered when developing anti-infective coatings.

Importantly, the inflammatory response to the surface properties of the biomaterial needs to be considered in all of its complexity. Specifically, it is important to look beyond the cellular and tissue morphology of the peri-implant region to consider the densities, activities and functions of other cells which may be impacted by endogenous and autacoid mediators (Anderson 2001; Anderson and Miller 1984). Moreover, while the histopathological consequences of acute inflammation are relatively uniform, the same cannot be said for chronic inflammation, where multiple factors can alter the histological appearance of the process. Enduring stimulations, e.g. those by a controlled-released antimicrobial agent, chemical or morphological surface features of the implant, may contribute to the establishment of chronic inflammation. The foreign body response is also influenced by the surface topography, configuration and the surface-to-volume ratio of the implant, with smoother implants having been associated with low levels of macrophages and foreign body giant cells in the peri-implant milieu and a tendency to develop a fibrous capsule. Rough, porous and fabric-like surfaces, on the other hand, have more macrophages and foreign body giant cells in the proximity to the implant, and are less likely to become encapsulated.

7.1.5 Evaluating Biocompatibility

Application of any type of physical or chemical modification to the surface of the implant will directly affect the dynamics of the above described host-implant relationships (Triantafillopoulos and Papaioannou 2014). It is therefore essential to determine whether or not the modified implant is still sufficiently biocompatible for the intended application. This can be done by assessing a series of biological responses, and then comparing the magnitude and duration of the adverse changes in homeostatic mechanisms of the host against those for the unmodified implant.

Consideration needs to be given to the nature and duration of the implant-tissue contact. For example, for medical devices that come into direct contact with blood, good hemocompatibility is paramount. Silver is an element with well-known antibacterial properties that has many promising applications, however, contradictory reports exist about the ability of silver nanoparticles to induce platelet aggregation and procoagulant activation, potentially leading to increased risk of cardiovascular events (Jun et al. 2011; Smock et al. 2014; Shrivastava et al. 2009; AshaRani et al. 2009). The effect of the implant on the mechanisms of thrombosis, coagulation, platelet activation, and production of blood and its components, i.e. blood cells, hemoglobin, blood proteins, will depend on the surface physico-chemical properties of the implant, as well as its geometry, contact conditions, and flow dynamics.

The contact duration is regarded as limited for contact time of less than 24 h, prolonged when implant-tissue contact is maintained for over 24 h to 1 month and permanent thereafter. The location of the implant and the types of tissues and bodily fluids that will be in direct contact with the implant are also considered.

For anti-infective modifications that involve biodegradation and controlled release of particles and/or chemical substances into the peri-implant milieu, the transport, accumulation, degradation and metabolism of these release by-products by other tissues may be of significance. Even very small amounts of metal ions, antimicrobials, biomolecules, nano- and micro-particles can give rise to allergic or sensitization reactions, with the severity of the reaction dependent on the properties of the leachable agent and the sensitivity of the host.

In cases where these biomaterials are implanted into susceptible individuals or are present in higher amounts, leachable agents can induce systemic toxicity, directly affecting the immune, hematologic, central nervous and cardiovascular systems, amongst others. In addition to leachable agents, any structural components of bacteria, yeasts, and molds, organic and inorganic dusts, nanoparticles, and diesel exhaust particles that may remain on the surface of the implant post-sterilization may provoke an inflammatory (innate) response and be fever-producing (pyrogenic) (Mazzotti et al. 2007).

Given that many modern antimicrobial and tissue regeneration strategies are based on the controlled release of biologically active molecules, including growth factors, cytokines, antimicrobial factors and proteins, stem cell stimulating factors, complement proteins, and chemotactic factors, it is important to consider the immunotoxic and immunogenic potential of these reactive agents (Maeda et al. 2007;

Bielecki et al. 2007; Chandorkar et al. 2015). Once released inside the host, these molecules, their fragments or their degradation by-products can be recognized by the adaptive immunity, triggering the humoral and cellular response. As such, exposure to these substances can lead to immunosuppression, immunostimulation, or autoimmunity, manifested as histopathological changes, host tissue autoimmune damage, and compromised ability of the host to protect itself from foreign agents. For example, an inorganic compound, protein or lipid component in the drug delivery system may perform as an adjuvant, stimulating autoimmune response; similarly, a foreign biomolecule may induce antibodies that will react with host molecules, leading undesirable tissue damage via complement pathway (Anderson 2001).

Systemic toxicity is typically evaluated as a function of exposure time, with relevant adverse reactions being observed within 24 h after first contact (acute), after 14–28 days of repeat exposure (sub-acute), and up to 90 days of exposure (sub-chronic). Chronic toxicity arises from long term exposure, and is characterized by persistent or progressively deteriorating dysfunction of cells, organs or multiple organ systems. Unlike other types of toxicity, it is difficult to evaluate and predict the chronic toxicity potential of the modified implant, due to the obvious difficulty in maintaining the experimental conditions, such as nutrition, health, and lifestyle parity between test groups.

Implant-related physical, chemical and biological agents may lead to genotoxicity. For example, particles and ions leached from the surface of metallic bone implants have been shown to increase DNA damage, gene mutations, and chromosome aberrations in the adjacent bone marrow and chromosome translocations and aneuploidy in the peripheral blood (Gajski et al. 2014; Karahalil et al. 2014). Such mutations, mistimed event activation, and direct DNA damage may lead to the development of cancer, although the risk of implant-leached products inducing systemic cancers is regarded as very low (Christian et al. 2014; Moalli et al. 2014). The carcinogenicity of the compound is patient-specific in that it depends on the ability of the individual to activate or detoxify genotoxic substances, and to repair deletions, breaks and/or rearrangements within DNA. Interestingly, biomaterial-associated tumor development in animals has been demonstrated to be associated with the physical configuration, not the chemical composition of the implant (Moalli et al. 2014). Implants with a smooth surface and large surface area, such as thin sheets and discs, are more carcinogenic than biomaterials with irregular surface morphologies, such as meshes and porous solids.

Implant-derived particles and reactive chemical species may affect reproductive toxicity. For instance, the reproductive potential of male mice injected with CoCr nanoparticles (to simulate leaching of particles and ions from CoCr alloy implants) was lower than that of the control group, with reduced epididymal sperm motility, viability and concentration, and higher abnormal sperm rate (Wang et al. 2013). Popular anti-microbial agents, silver and titanium dioxide nanoparticles were also found to be cytotoxic and cytostatic for primary testicular cells, causing apoptosis, necrosis and decreased proliferation (Asare et al. 2012). Oxidative stress is the likely driver for the testicular damage and pathological changes. Indeed, the production of reactive oxygen species and DNA damage are regarded as the underlying mechanism of nanoparticle cytotoxicity (Taylor et al. 2012). Compared to bulk materials, nanoparticles have heightened chemical and biological reactivity due to much higher surface-to-volume ratio, which means that even inert biocompatible materials, e.g. gold (Zhao et al. 2013b), can become toxic. Specific toxicity is highly dependent on the size, shape, polarization and surface functionalization of the particle.

Nanoparticles may induce teratogenicity, halting the development or leading to congenital malformation of the fetus, and negatively affecting post-natal development of the offspring. For example, graphene oxide is a promising antimicrobial material (Liu et al. 2011; Podila et al. 2013), which, when modified with silver nanoparticles, becomes even more strongly antibacterial towards pathogens such as *E. coli* (Ma et al. 2011). When maternal mice were exposed to graphene oxide during lactation, the filial mice developed significantly slower, gaining less body weight and length compared to control group (Fu et al. 2015). The intestinal villus of the filial mice was found to be notably shorter in the offspring mice in the treatment group, indicating a possible mechanism of toxicity.

A critical determinant of toxicity is the dose to which the host is exposed. Biodegradable and drug-release antimicrobial coatings are designed to release tightly controlled quantities of biochemically active substances into the peri-implant milieu. The release rate balances microbial efficacy against host toxicity. Premature biodegradation *in vivo*, i.e. excessive elimination of the therapeutic compound, may tip that balance. In addition to toxicity, distorted degradation kinetics may result in the premature exhaustion of the anti-bacterial agent. The suboptimal levels of the released agent may not be sufficient to effectively kill bacteria, potentially acting as a positively driver for the development of antibiotic resistivity.

In addition to the active ingredient, toxicity can arise from other seemingly benign degradation products, including very low amounts of impurities, additives and chemicals used in the synthesis and processing of a biomaterial. The chemical and morphological properties of the released agents will be affected by the biodegradation mechanism and on the health status of the host (Baran et al. 2014). Medically compromising systemic conditions, e.g. diabetes, and inflammatory diseases may increase the rate of biomaterial degradation (Alani and Bishop 2014). Higher blood glucose levels and lower pH levels, for instance, has been shown to the rate of electrochemical corrosion of titanium dental implants (Tamam and Turkyilmaz 2014). It is challenging to predict all the potential interactions among the degradation products, and also between these products and other biochemical agents used for the treatment of the patient. One of the issues is associated with the difficulty in identifying appropriate biomarkers for detection of premature degradation onset and progression (Sumner et al. 2014; Grainger et al. 2013).

7.2 Modern Strategies to Minimize Implant-Associated Infections

There exists a wide spectrum of substances and technological methodologies that can be used to fabricate biomaterials with anti-infective features (Jadalannagari et al. 2014), including those based on bio-inspired and biomimetic surface topographies (discussed in Chap. 2), electroactive and mechano-responsive surfaces (Chaps. 4 and 8), and natural antimicrobial agents (Chap. 2). Broadly, anti-infective approaches can be divided into two categories: antifouling, i.e. those that repel microbes; and antimicrobial, those that prevent cell proliferation (bacteriostatic) and/or kill bacteria (bactericidal). In the following sections, we will discuss several examples of the modern strategies to minimize implant-associated infections, focusing on the challenges in merging the need to withstand bacterial load with that to support function restoration and tissue healing.

7.2.1 Anti-adhesive Surfaces

Device related infections take place when bacteria coat the surface or infiltrate the lumen of an implant. The peri-implant milieu is abundant in nutrients, providing suitable growth conditions for bacterial growth and replication. On the surfaces of implants, bacteria can be present in both planktonic and biofilm forms, the former affording mobility whereas the latter providing means of protection, selective transport of biomolecules, environmental regulation and communication for resident cells. The presence of biofilm markedly increases the likelihood of bacteria evading host defense mechanisms, antimicrobial agents and extended periods of unfavorable environmental conditions. The release of extracellular polymeric substances, e.g. polysaccharides, amyloid fibrils, lipids and nucleic acid (Claessen et al. 2014), is critical to establishment, development and functioning of any biofilm. For example, adhesins enable cell attachment to living tissue or an abiotic substrate and act as receptors for adhesins on other bacteria. A detailed overview of the process of bacterial attachment and biofilm formation can be found in Chap. 1.

Since adherence is a critical first step in pathogenesis, many strategies have been developed to control bacterial and protein adhesion via steric repulsion, low surface energy and electrostatic repulsion. This can be achieved by modifying the surface to be hydrophilic, negatively charged, and/or with low surface free energy (Yu et al. 2011). In the case of hydrophilic polymers, water is attracted to the surface, encapsulating it in a thin repellent layer that is associated with the polymer via hydrogen bonds. This layer creates steric barrier for the biomolecules and cells, preventing their attachment.

The appropriate chemistry is typically obtained by surface functionalization, e.g. plasma treatment (Bazaka et al. 2011, 2012; Jacobs et al. 2012), or application of coatings and polymer brushes (Knetsch and Koole 2011; Hook et al. 2012; Chauhan et al. 2014). Hydrophilicity of the material may also be altered by controlling surface morphology (Zheng et al. 2010). For example, metallic surfaces can be made superhydrophobic using chemical etching or laser ablation (Fadeeva et al. 2011). Amongst polymer brushes, amphiphilic copolymers (Liu et al. 2012b; Zhou et al. 2014), zwitterionic polymers (Cheng et al. 2008) and patterned polymers are common. Polymer chemistries based on polyethylene glycol (PEG) or its derivatives are common (Ye and Zhou 2015; Yang et al. 2014), as they are non-immunogenic and effectively resist protein fouling and thrombosis (Chen et al. 2011). The efficacy of PEG modification is strongly dependent on the molecular weight, branching and surface chemistry of the coating (Beloin et al. 2014). In addition to PEG, hydrophilic polymers based on hyaluronic acid (Liu et al. 2014) such as poly-N-vinylpyrrolidone (Liu et al. 2013a), poly(dimethylaminoethyl methacrylate) (Lih et al.; Tu et al. 2013), have also been used.

There are many possible configurations in which polymer brushes can be assembled, and enumerate chemistries that can be used to fine tune their properties for a desired application or to enhance the antifouling property of the surface. Broadly, polymer can be attached in the form of a homopolymer, as mixed polymer or block co-polymer brushes, assembled from functional particles or as layer-by-layer films, incorporating nano- an micro-sized particles, in a form of cross-linked polymer matrixes, and so on (Lih et al.; Wang et al. 2014). For example, the antifouling efficacy of fluorinated amphiphilic copolymers composed of 2-perfluorooctylethyl methacrylate and 2-hydroxyethyl methacrylate was greater than that of their constituent homopolymers, suggesting a synergistic mechanism of fouling inhibition (Zhao et al. 2013c). The stability of the polymer layer is dependent on the method of immobilization, with physically absorbed (self-assembled) brushes being less chemically stable than covalently linked molecules, e.g. grafted-to or grafted-from polymer chains.

Improved anti-infective outcomes can be achieved by synergistically merging antifouling and antibacterial chemistries, e.g. by coating the implant surface with brush-like polycarbonates containing pendent adhesive dopamine, antifouling PEG, and antibacterial cations (Yang et al. 2014). Such a surface can ensure that bacterial cells able to overcome antifouling chemistry and attach to the implant surface are prevented from multiplying and forming a biofilm (Yang et al. 2014). In another example, antifouling poly(sulfobetaine methacrylate) and bactericidal polymer, *N*-[(2-hydroxy-3-trimethylammonium)propyl] chitosan chloride polymers are integrated to achieved a similar result (Wang et al. 2015). Heparin and other biomolecules can be loaded into such a coating to enhance surface hemocompatibility and improve host cell attachment (Almodóvar et al. 2013; Zhu et al. 2013). Coating the surface of the implant with heparin and heparin-like molecules has been demonstrated to reduce bacterial attachment and biofilm formation. Presence of heparin or albumin results in a negatively charged surface that reduces bacterial adhesion (as cells bear a net negative charge on their surface under normal pH conditions) and the deposition of fibronectins (Francolini and Donelli 2010).

From a biocompatibility point of view, most anti-fouling surfaces are non-toxic, as the effect is typically achieved by surface chemistry rather than a release of antimicrobial agents. At the same time, the antifouling action is generally non-specific, i.e. it may alter the attachment of host biomolecules and cells. While this may be desired for urinary or vascular catheter applications or wound dressings, it may impede the establishment of a tissue-implant interface where integration is desired. To circumvent this problem, antifouling polymer brushes can be functionalized or loaded with biomolecules that promote host cell adhesion, tissue formation and spreading. For example, improved tissue compatibility was obtained by functionalizing brushes of the block copolymer Pluronic F-127 with contact-killing antimicrobial peptides and host cell adhesion promoting arginine-glycine-aspartate (Muszanska et al. 2014) (Figs. 7.1 and 7.2).

Higher control over host cell adhesion can be obtained by specifically targeting bacterial adhesion, by using strategies that hinder biogenesis of fimbrial adhesins (Lo et al. 2013; Nait Chabane et al. 2014; Shamir et al. 2010). Production of functional extracellular amyloid fibers by uropathogenic *E. coli* and other Enterobacteriaceae can be effectively hindered *in vitro* and *in vivo* by using ring-fused 2-pyridones, such as FN075 and BibC6 (Cegelski et al. 2009). Pyridone-based pilicides inhibit the assembly of type 1 pili and thus interfere with pathogenesis and formation of curli-dependent and type 1-dependent biofilms. Another target is the FimH, an adhesion located at the tip of type 1 fimbriae and responsible for binding to mannosylated glycoproteins on human cells (Totsika et al. 2013). Here, small molecule mannosides can be used to bind FimH thus making it unavailable for binding to eukaryotic receptors (Guiton et al. 2012). Glycoclusters based on a cyclic oligo-(1→6)-β-D-glucosamine core can be employed to inhibit bacterial lectin LecA of *P. aeruginosa* (Gening et al. 2013), interfering with the colonization ability

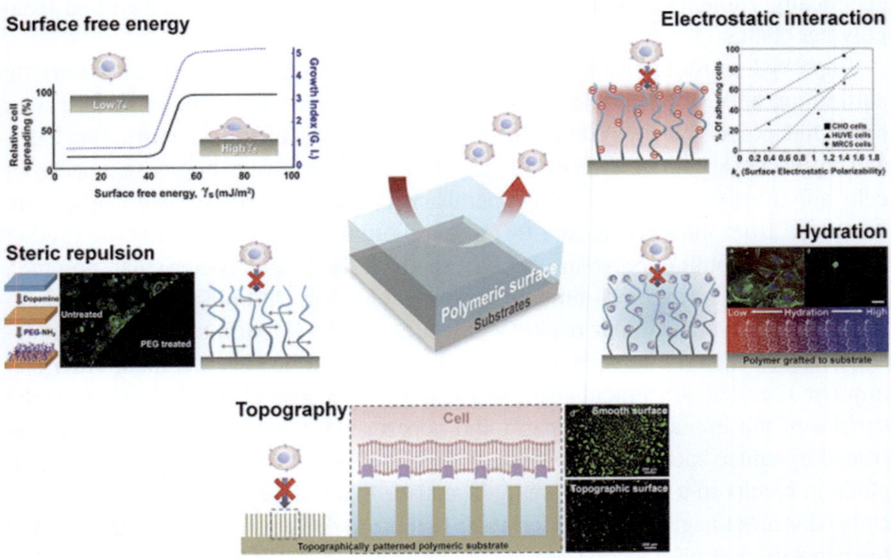

Fig. 7.1 Interactions between host cells and anti-adhesive surfaces (Reproduced with a permission (Lih et al.))

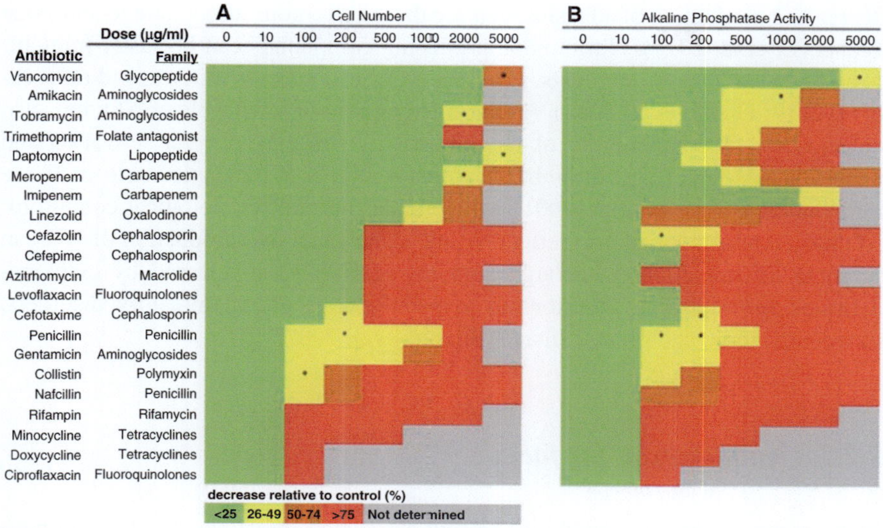

Fig. 7.2 Effect of various concentrations of antibiotics on osteogenic cell viability and activity. The mean % decreases in osteoblast cell number (**a**) and ALP activity (**b**) are classified as <25 %, 26–50, 51–74, and >75 % of control after incubation with 0, 10, 100, 200, 500, 1000, 2000, and 5000 µg/mL of each antibiotic for 10 and 14 days (n = 5–6 per dose after data are pooled). Not determined: ALP activity and/or cell number was untestable for some of the antibiotics due to precipitation or incompatibility with the test assays used. Decreases in osteoblast cell number and ALP activity >25 % were significant, $p < 0.05$, with exceptions indicated by (*) where the value at that dose was not different from control (Reproduced with permission (Rathbone et al. 2011))

of the pathogen. Furthermore, the biofilm-forming ability of *P. aeruginosa* can be compromised by targeting several specific functions of the pathogen. For example, maltose derivatives with bulky hydro-carbon groups inhibit the swarming motility and cell adhesion of *P. aeruginosa*, demonstrating biofilm inhibition and dispersion (Shetye et al. 2014).

Another potential target is small messenger molecules that enable pathogen cell-to-cell communication within a colony. Being a co-operative process, biofilm formation is inherently dependent on information being effectively transported from cell to cell. Quorum sensing is responsible for synchronization of gene expression, co-ordinated switch between planktonic and sessile states, biofilm formation, maturation and disassembly (Solano et al. 2014; Singh et al. 2000). By targeting the major signaling pathways of cells, bacterial communication and thus biofilm formation can be compromised. Sulfathiazole and azathioprine has been shown to inhibit diguanylate cyclases (Sambanthamoorthy et al. 2012, 2013; Antoniani et al. 2010, 2013). This enzyme is responsible for production of c-di-GMP, a small messenger molecule responsible for switching from planktonic to biofilm modality.

Nitric oxide is another molecule that can interfere with biofilm formation and induce dispersal of *P. aeruginosa* (Barraud et al. 2009). Nitric oxide increases cell motility by decreasing the levels of the secondary messenger cyclic di-GMP and decreasing production of adhesins, making these cells more susceptible to antibiotic agents (Li et al. 2013). Nitric-oxide producing compounds can be easily immobilized on the surfaces of implants, releasing NO under specific *in vivo* conditions (Li et al. 2013; Duong et al. 2014). A controlled release of low levels of NO may also enhance the biocompatibility of the surfaces by reducing platelet activation and adhesion, and inhibiting thrombus formation. NO has a complex role in tissue inflammation (Laroux et al. 2001), where under normal physiological conditions, NO has an anti-inflammatory effect. At the same time, overproduction of NO can promote inflammation, with large amounts of NO being released by cytokine-activated macrophages, contributing to pathogenesis of inflammatory disorders, vasoconstriction, and tissue damage (Sharma et al. 2007).

7.2.2 Antibacterial Surfaces

Most strategies aimed at reducing biofilm-associated infections combine an antifouling property with antimicrobial activity (Francolini et al. 2015; Salwiczek et al. 2014), either through elution of antimicrobial agent or a surface killing. Antimicrobial agents range from conventional (systemic) antibiotics (Gao et al. 2011), e.g. nafcillin, levofloxacin, daptomycin, gentamycin, vancomycin (Bastari et al. 2014; Beenken et al. 2014; Ordikhani et al. 2014; Wu et al. 2014; Cashman et al. 2013), to metallic ions such as silver (Wong and Liu 2010; Chernousova and Epple 2013; Perez et al. 2014; Jo et al. 2014) and copper (Kalaivani et al. 2014; Ye et al. 2014; Chen et al. 2014), nitrofurazone (Kottur et al. 2015; Johnson et al. 2012), chlorhexidine (Jamal et al. 2014), quaternary ammonium compounds (Asri et al. 2014; Shalev et al. 2012; Cheng et al. 2012; Bakhshi et al. 2013), antibacterial peptides (Muszanska et al. 2014; Cleophas et al. 2014; Rapsch et al. 2013; Etayash et al. 2013) and anionic nanoporous hydrogels (Li et al. 2011; Hook et al. 2012). These agents are highly diverse in their modes of action. For example, an antimicrobial polycationic hydrogel based on dimethyldecylammonium chitosan-graft-PEG methacrylate and PEG diacrylate attracts sections of the anionic membrane of the pathogen into the internal nanopores of the hydrogel (Li et al. 2011). This process leads to bacterial membrane disruption and subsequent death. Synthetic macromolecular antimicrobials are being researched for their potential to physically destroy cell membranes of the pathogens thus preventing them from developing drug-resistance. Examples of this type of antibacterials include biodegradable cationic polycarbonates containing propyl and hexyl side chains quaternized with various nitrogen-containing heterocycles, such as imidazoles and pyridines (Ng et al. 2014). Although not based on a particular chemistry, surface capable of contact killing using morphological features are also being investigated (Ivanova et al. 2012, 2013).

7.2.2.1 Systemic Antibiotics

Conventional antibiotics have the advantage of having been thoroughly investigated, with well-defined host toxicity profiles and histories of clinical use that might detail potential long-term side effects. Since the doses of antibiotic released from the surface of implant are typically notably smaller than those used systemically, the toxicity of drug-eluting coatings is considered to be low. Indeed, one of the key advantages of drug-release coatings is the ability to locally deliver relatively high doses of antibiotic without inducing systemic toxicity. Nonetheless, the effects of antibiotics on cell viability and tissue regeneration should be considered. Osteoblasts treated with 21 different antibiotics over 0–5000 μg/mL concentrations for up to 14 days showed significantly lower cell number and osteogenic activity when exposed to rifampin, minocycline, doxycycline, nafcillin, penicillin, ciprofloxacin, colistin methanesulfonate, and gentamicin (Rathbone et al. 2011). On the other hand, osteoblast deoxyribonucleic acid content and alkaline phosphatase activity were least affected by amikacin, tobramycin, and vancomycin (Rathbone et al. 2011). Nevertheless, the majority of the antibiotics tested, a ≥50 % decrease in osteoblast cell number and/or osteogenic activity was observed. The decrease in the metabolic activity and thus the osteogenic potential of surviving cells may undermine bone repair.

7.2.2.2 Metals Salts

Metallic salts, such as those of silver, copper, zinc and mercury have been used to prevent and treat infections in wounds and burns for thousands of years. Long before the invention of polymer sutures, silver threads were also used to close wounds (Muffly et al. 2011), while silver-based solutions and creams (e.g. silver nitrate, silver sulfadiazine) were used as washes and ointments. Silver-based compounds precipitate cellular proteins and interfere with respiration of both aerobic and anaerobic bacteria, however the exact mechanism is yet to be fully elucidated (Chernousova and Epple 2013). Silver can also be used as part of combination therapy, where its ability to damage bacterial cells is used to potentiate antibiotic activity of conventional antibiotics (Morones-Ramirez et al. 2013) and thus target persister cells.

Current uses of silver primarily focus on a nano-sized particulate form of the metal (Rai et al. 2009; Wong and Liu 2010; Taheri et al. 2014). The nanoparticle form is characterized by high surface-to-volume ratio, which makes the particulate form of silver significantly more active with regard to its physicochemical and biological properties. It is believed that large surface area of nanoparticles allows them to establish good contact with the surfaces of target cell, whereas small size may facilitate particle penetration inside the pathogen. Larger surface area also enables more efficient release of silver ions, which can then interact with sulfur- and phosphorous-containing biomolecules in the bacterial membrane and DNA, respectively. Enhanced chemical reactivity of silver nanoparticles leads to the generation of a large number of reactive oxygen species, which inflict further damage to the cells.

From a biocompatibility and cytotoxicity point of view, the major issue with the use of metallic ions is that their biological interactions are non-specific. Indeed, silver nanoparticles can easily breach host tissue and cell barriers, and interact with host biomolecules, e.g. enzymes, altering cell signalling and metabolic pathways (Gérard et al. 2010; Taylor 1985; Mouriño et al. 2012). Composition, dimensions, configuration and functionalization will affect the transport of chemical reactivity, and as such, the antibacterial efficacy and the potential toxicity of the nanoparticles (Li et al. 2012; Rai et al. 2009). For example, when human bone marrow mesenchymal stem cells and human hepatoma carcinoma cells were incubated in the presence of gold nanoparticles of different size, survival was 80 % in the case of 15 and 30 nm cells and less than 60 % for 5 nm particles (Fan et al. 2009). The increased levels of cell necrosis and altered osteogenic and adipogenic capabilities were attributed to the increase in the level of reactive oxygen species.

Anti-proliferative activity of silver nanoparticles was demonstrated in human lung fibroblasts (IMR-90) and glioblastoma cells, with treated cells exhibiting chromosome instability and mitotic arrest (AshaRani et al. 2009). Exposure to silver nanoparticles induced apoptosis in NIH3T3 fibroblast cells, inducing the release of cytochrome c into the cytosol and translocation of Bax to mitochondria through ROS and JNK (Hsin et al. 2008). Cellular apoptosis was also induced in RAW264.7 macrophage cells cultured in the presence of ~70 nm silver nanoparticles (Park et al. 2010). These particles were found to lower the intracellular glutathione level, increase NO secretion and TNF-α in protein and gene levels, and increase the gene expression of matrix metalloproteinases. It was demonstrated that intracellular oxidative stress due to the presence of agglomerated nanoparticles rather than ion damage was responsible for apoptosis in human hepatoma HepG2 cells (Kim et al. 2009). DNA damage may have also played a role, as hepatoma cells cultured with low dose silver nanoparticles showed an upregulation of DNA repair-associated genes (Kawata et al. 2009).

Oxidative stress was determined to be the cause of HeLa cell death, where oxidative stress-related genes, ho-1 and mt-2A, were expressed (Miura and Shinohara 2009). Coated silver nanoparticles also induced oxidative stress-type damage onto N27 rat dopaminergic neurons, where the size of the particle and type of coating determined the dominant pathway (Chorley et al. 2014). Poly(vinyl pyrrolidone)-coated particles were found to be more bioactive than citrate-coated particles, increasing intra-neuronal nitrite levels and inducing mitochondrial dysfunction (75 nm particles) and NRF2 oxidative stress pathway (10 nm particles) (Veronesi et al. 2014).

In the case of copper particles, the physical form of the particle is important, with dissolved Cu^{2+} ions contributing <50 % to cytotoxicity (Wang et al. 2012). A comparison between cell-particle interactions of nano-sized CuO and Ag showed that in the case of CuO, the particle was subject to rapid uptake by endocytosis, releasing copious amounts of copper ionic species within the cells (Cronholm et al. 2013). Even though silver particles were readily taken up by the cells, the intracellular release of silver ions and hence the toxicity of the particles was low. The mechanism of toxicity also differed between different types of copper, where nanoparticles of copper and Cu-Zn alloy inflicted substantial damage onto cell membranes, whereas this behavior was not observed in the case of CuO nanoparticles and the micron-sized Cu metal particles (Karlsson et al. 2013). Protein fouling in the presence of serum albumin inhibited cell toxicity of silver nanoparticles (Gnanadhas et al. 2013).

The systemic toxicity of medically-relevant metallic nanoparticles is not well studied (Alenius et al. 2014). Once eluted from the surface of the implant, the particles should theoretically be able to quickly and easily cross the blood barrier, being transported to other tissues in the body. There, these nanoparticles can potentially accumulate and degrade (Wong and Liu 2010). With regard to the cardiovascular system, sufficiently high doses of nanoparticles can alter the microcirculation, and promote thrombus formation and pro-inflammatory responses (Wong and Liu 2010), whilst in the central nervous system metallic nanoparticles may generate oxygen reactive species and damage the brain cells (Alenius et al. 2014). An iron- and oxygen-binding protein found in the muscle tissue, myoglobin can be affected by the presence of silver nanoparticles, where ~10 nm particles were shown to significantly augment its electron-transfer reactivity and catalytic ability toward hydrogen peroxide (Gan et al. 2004). They may also affect the reproductive ability of the host, as silver nanoparticles showed concentration dependent cytotoxicity towards mouse spermatogonial stem cells (Braydich-Stolle et al. 2005).

7.2.2.3 Chlorhexidine and Quaternary Ammonium Compounds

Chlorhexidine, a bisbiguanide, is an effective hospital disinfectant often used in surgery skin preparation and as an antimicrobial agent in medical devices (Opstrup et al. 2014; Silvestri and McEnery-Stonelake 2013) where they are employed to reduce extraluminal contamination, particularly in patients who require long-term vascular access (Khoo and Oziemski 2011). Recently, concerns have been raised about potentially significant allergic reactions to chlorhexidine-impregnated medical devices (Guleri et al. 2012). Severe allergic reactions, including anaphylactic shock have been reported for patients during the placement of a chlorhexidine impregnated central venous catheters (Khoo and Oziemski 2011; Faber et al. 2012; Bae et al. 2008), urological and rectal procedures (Jayathillake et al.; Bae et al. 2008). Approximately 10 % of patients with suspected perioperative allergic reactions were diagnosed with chlorhexidine (Opstrup et al. 2014). Wide exposure to chlorhexidine in hospital and non-medical environments may potentially sensitize some patients, leading to adverse reactions during surgery.

Quaternary ammonium compounds (QACs) are potent cationic antimicrobials active against a wide range of pathogens (Hegstad et al. 2010). QACs kill bacteria by interfering with cell membranes, primarily the cytoplasmic membrane, leading to membrane damage and loss of cellular content (Martín et al. 2014). QACs block potassium channels that are responsible for the passive movement of potassium ions across the cell membrane and control the membrane potential in both eukaryotic and prokaryotic cells (Lenaeus et al. 2014). QACs may disrupt and denature structural proteins and enzymes. In anti-infective device applications, QACs can be loaded into coatings or surface immobilize for contact killing (Asri et al. 2014). As with other non-specific antimicrobial compounds, quaternary ammonium compounds display concentration-dependent eukaryotic cell toxicity with all types of exposure, namely inhalation, ingestion, and dermal application and irrigation of body cavities (Xue et al. 2004, 2012; Świercz et al. 2013; Dutot et al. 2008a, b). In susceptible individuals, contact with QACs may result in allergies that can range from mild localised irritation to anaphylactic reaction (Bello et al. 2009) and contact tissue damage (Kilic et al.). Mucosal contact with QACs may produce sensitization.

7.2.2.4 Antimicrobial Peptides

Produced by a large number of microorganisms, plants, invertebrates and animal species, antimicrobial peptides fulfil a broad range of roles in producing an innate immune response, specifically they hinder the growth and colonisation by infectious agents (Zasloff 2002; Bals et al. 1999). For example, resident epithelial cells (e.g. keratinocytes) produce a number of antimicrobial peptides and also signal the recruitment of circulating immune cells (e.g. neutrophils); once summoned, neutrophils also produce antimicrobial peptides, such as cathelicidin and β-defensins to fight the infection (Braff et al. 2005; Nizet et al. 2001). When adsorbed onto surfaces, nisin, an antimicrobial peptide, inhibits the attachment and biofilm formation capability of Gram-positive bacteria (Bower et al. 2002), whereas covalently-bound to the surface cathelin LL37 effectively kills *E. coli* on contact (Zaiou et al. 2003).

In Gram-positive and Gram-negative organisms, the antibacterial mechanism of the antimicrobial peptides is biophysical rather than biochemical in nature, destabilizing and disorganizing the structure of the membrane (Fitzgerald-Hughes et al. 2012). The cationic nature and amphipathic structure of antimicrobial peptides allows them to electrostatically attach to the negatively charged microbial surfaces (Powers and Hancock 2003; Forbes et al. 2013; Zasloff 2002). Subsequently, some peptides induce pore formation in the phospholipid bilayer, with death being a consequence of osmolysis (Augustyniak et al. 2012). A compromised cytoplasmic membrane also enables the translocated peptides to interact with intracellular biomolecules of the pathogen, inhibiting protein synthesis and conformation, interfering with DNA and metabolic activity, and suppressing cell multiplication (Brogden 2005). For example, the antibacterial action of nisin involves bacterial membrane permeation and disruption of the ability for the cell wall to undergo synthesis and regeneration (Hale and Hancock 2007).

Given that antimicrobial peptides typically combine several mechanisms of action, they are less likely to prompt de novo resistance (Yeung et al. 2011). The killing efficacy, and potentially the antibacterial mechanism itself varies with peptide structure and the target organism (Augustyniak et al. 2012; Veiga et al. 2012;

Salick et al. 2007, 2009; Liu et al. 2013b; Zhou et al. 2011); however, there is only a limited understanding of all the mechanisms involved, which limits the optimization and clinical use of these peptides (Sahl et al. 2005). Further hindrance comes from the *in vivo* instability of these antimicrobial agents, whereby they are easily degraded by host enzymes, in addition to being affected by pH and osmotic conditions (Brogden and Brogden 2011). Various structural modifications have been proposed to improve peptide resistance to enzymatic proteolysis without compromising its killing efficacy (Zasloff 2002; Hamuro et al. 1999; Porter et al. 2000).

Being a foreign protein, host defense peptides can potentially be immunogenic. Host defense peptides are considered to be weakly immunogenic, owing to their small size and linear structure (Fitzgerald-Hughes et al. 2012). Nevertheless, antibodies were produced to human neutrophil defensins (Panyutich et al. 1991), bovine lactoferricin(R) (Shimazaki et al. 1996) and hCAP-18 (Sørensen et al. 1997). Furthermore, it is unclear whether the use of foreign peptides will lead to the development of peptide resistance in pathogenic bacteria, and thus compromise the natural ability of the host to fight infection (Fitzgerald-Hughes et al. 2012). The effect of synthetic human defense peptide-like molecules on the natural innate response has yet to be studied.

With regard to selectivity and cytotoxicity of antimicrobial peptides (Hancock and Sahl 2006), the comparative absence of negatively charged lipids on the surfaces of eukaryotic cells and their weak membrane potential gradient may afford some protection to host cells (Fitzgerald-Hughes et al. 2012). Some host defense peptides, e.g. magainin 2 and human cathelicidin LL-37, can translocate into the cytosol of mammalian cells, e.g. HeLa, TM12 and Chinese hamster ovary (CHO)-K1 cells (Imura et al. 2008; Takeshima et al. 2003). At higher concentrations of the magainin 2 peptide, cell toxicity was observed. The translocation across host cell is possible owing to a dual role of peptides such as LL-37, i.e. bacterial targeting and delivery of nucleic acids into the host cells (Zhang et al. 2010). Even though LL-37 is not haemolytic at antibacterial-relevant concentrations, *in vitro* cytotoxicity has been observed. Toxicity towards eukaryotic cells of orangutan, rhesus macaque and leaf eater monkey orthologues of LL-37 varied with the physical characteristics, with the leaf eater monkey peptide being most toxic (Tomasinsig et al. 2009). Cytotoxicity also depended on the nature and metabolic state of the target cells. At the same time, haemolytic activity was found to be similar among the tested peptides.

The preferential activity against Gram-positive and/or Gram-negative bacteria membranes over eukaryotic cells can be enhanced by controlling the ratio of D- to L-amino acids (Fernandez-Lopez et al. 2001; Oren and Shai 2000; Yin et al. 2012). Structural modifications, e.g. truncation or rearrangement, can be used to enhance membranolytic efficiency (Shimizu et al. 1998). If it is done while disregarding other relevant targets, however, the increased non-specific membranolytic activity can undermine surface bio-compatibility to the extent that this modification will have limited *in vivo* use. Target-specific antimicrobial peptides, where a *Pseudomonas*-specific targeting moiety is appended to a generally killing peptide novispirin G10, can both improve the speed and efficiency of bactericidal action (Eckert et al. 2006).

7.3 Conclusion

The host response to the act of implantation initiates a highly complex and multi-dimensional cascade of events, that start with an injury and proceeds to include blood-surface interactions, provisional matrix formation, acute and chronic inflammation, foreign body reaction, development of granulated tissue, healing and fibrous capsule formation. The exact events taking place will depend on a multitude of different parameters, including those pertinent to the host, those associated with the quality of the surgical procedure, and those that result from interactions between living host tissues and the surface and bulk characteristics of the biomaterial. Not surprisingly, predicting or even describing the biocompatibility of an implant is not trivial. The task becomes even more challenging when biodegradable and eluting surfaces are considered. It is these surfaces, however, that hold the greatest promise in mitigating biomaterial-associated infections. Indeed, where antifouling surfaces may succeed in some short term applications, long-term and permanent devices require a reliable means to ensure pathogen-free surfaces. Surface modification that synergistically combines multiple complementary lines of defense may provide the most effective and durable solution. Nevertheless, balancing the biocompatibility of an implant against the ability of their surfaces to be efficient over a range of bacterial loads is a difficult task. With the rise of alternative antimicrobial agents or configurations, more information is needed regarding their compatibility, toxicity and immunogenicity towards relevant host cells. In order to attain a more comprehensive picture, it is important to approximate the conditions under which implants will operate and the quantities of active agents that can potentially be leached out into the peri-implant milieu. A standardized experimental procedure with a focus on both the anti-infective efficacy and biocompatibility should provide for a better comparison between different antimicrobial strategies. It should also provide valuable information on which material and process optimization can be based to ensure appropriate cell integration; after all, for many applications, the ultimate aim is to achieve an adequate tissue integration and a functional implant.

References

Abrahamsson I, Berglundh T, Linder E, Lang NP, Lindhe J (2004) Early bone formation adjacent to rough and turned endosseous implant surfaces. Frühe Knochenbildung bei rauen und gedrehten enossalen Implantatoberflächen Eine experimentelle Studie an Hunden. Clin Oral Implants Res 15(4):381–392. doi:10.1111/j.1600-0501.2004.01082.x

Alani A, Bishop K (2014) Peri-implantitis. Part 2: Prevention and maintenance of peri-implant health. Br Dent J 217(6):289–297. doi:10.1038/sj.bdj.2014.809

Alenius H, Catalán J, Lindberg H, Norppa H, Palomäki J, Savolainen K (2014) Chapter 3 – Nanomaterials and human health. In: Berges UVSWTB (ed) Handbook of nanosafety. Academic, San Diego, pp 59–133. http://dx.doi.org/10.1016/B978-0-12-416604-2.00003-2

Almodóvar J, Mower J, Banerjee A, Sarkar AK, Ehrhart NP, Kipper MJ (2013) Chitosan-heparin polyelectrolyte multilayers on cortical bone: periosteum-mimetic, cytophilic, antibacterial coatings. Biotechnol Bioeng 110(2):609–618. doi:10.1002/bit.24710

Amoroso A, Boudet J, Berzigotti S, Duval V, Teller N, Mengin-Lecreulx D, Luxen A, Simorre J-P, Joris B (2012) A peptidoglycan fragment triggers β-lactam resistance in <italic>Bacillus licheniformis</italic> PLoS Pathog 8(3):e1002571. doi:10.1371/journal.ppat.1002571

Anderson JM (2001) Biological responses to materials. Annu Rev Mater Res 31(1):81–110. doi:10.1146/annurev.matsci.31.1.81

Anderson JM, Miller KM (1984) Biomaterial biocompatibility and the macrophage. Biomaterials 5(1):5–10. http://dx.doi.org/10.1016/0142-9612(84)90060-7

Andreas H, Cornelia P, Uwe W, Silke L, Maciej P, Lutz W, Hans-Georg N, Michael S (2013) In vivo evaluation of copper release and acute local tissue reactions after implantation of copper-coated titanium implants in rats. Biomed Mater 8(3):035009

Anitua E, Prado R, Orive G, Tejero R (2014) Effects of calcium-modified titanium implant surfaces on platelet activation, clot formation, and osseointegration. J Biomed Mater Res A. doi:10.1002/jbm.a.35240

Antoniani D, Bocci P, Maciąg A, Raffaelli N, Landini P (2010) Monitoring of diguanylate cyclase activity and of cyclic-di-GMP biosynthesis by whole-cell assays suitable for high-throughput screening of biofilm inhibitors. Appl Microbiol Biotechnol 85(4):1095–1104. doi:10.1007/s00253-009-2199-x

Antoniani D, Rossi E, Rinaldo S, Bocci P, Lolicato M, Paiardini A, Raffaelli N, Cutruzzolà F, Landini P (2013) The immunosuppressive drug azathioprine inhibits biosynthesis of the bacterial signal molecule cyclic-di-GMP by interfering with intracellular nucleotide pool availability. Appl Microbiol Biotechnol 97(16):7325–7336. doi:10.1007/s00253-013-4875-0

Asare N, Instanes C, Sandberg WJ, Refsnes M, Schwarze P, Kruszewski M, Brunborg G (2012) Cytotoxic and genotoxic effects of silver nanoparticles in testicular cells. Toxicology 291(1–3):65–72. http://dx.doi.org/10.1016/j.tox.2011.10.022

Asharani P, Hande MP, Valiyaveettil S (2009) Anti-proliferative activity of silver nanoparticles. BMC Cell Biol 10(1):65

Asri LATW, Crismaru M, Roest S, Chen Y, Ivashenko O, Rudolf P, Tiller JC, van der Mei HC, Loontjens TJA, Busscher HJ (2014) A shape-adaptive, antibacterial-coating of immobilized quaternary-ammonium compounds tethered on hyperbranched polyurea and its mechanism of action. Adv Funct Mater 24(3):346–355. doi:10.1002/adfm.201301686

Augustine S (2014) All surgical site infections are not equal. Anesth Analg 118(4):883 810.1213/ANE.0000000000000104

Augustyniak D, Nowak JT, Lundy F (2012) Direct and indirect antimicrobial activities of neuropeptides and their therapeutic potential. Curr Protein Pept Sci 13(8):723–738. doi:10.2174/138920312804871139

Bae Y-J, Park CS, Lee JK, Jeong E, Kim T-B, Cho YS, Moon H-B (2008) A case of anaphylaxis to chlorhexidine during digital rectal examination. J Korean Med Sci 23(3):526–528

Bakhshi H, Yeganeh H, Mehdipour-Ataei S, Shokrgozar MA, Yari A, Saeedi-Eslami SN (2013) Synthesis and characterization of antibacterial polyurethane coatings from quaternary ammonium salts functionalized soybean oil based polyols. Mater Sci Eng C 33(1):153–164. http://dx.doi.org/10.1016/j.msec.2012.08.023

Balaban NQ, Merrin J, Chait R, Kowalik L, Leibler S (2004) Bacterial persistence as a phenotypic switch. Science 305(5690):1622–1625. doi:10.1126/science.1099390

Bals R, Weiner DJ, Moscioni AD, Meegalla RL, Wilson JM (1999) Augmentation of innate host defense by expression of a cathelicidin antimicrobial peptide. Infect Immun 67(11):6084–6089

Baran G, Kiani M, Samuel S (2014) Properties of the host (the human body). In: Healthcare and biomedical technology in the 21st century. Springer, New York, pp 183–202. doi:10.1007/978-1-4614-8541-4_6

Barraud N, Schleheck D, Klebensberger J, Webb JS, Hassett DJ, Rice SA, Kjelleberg S (2009) Nitric oxide signaling in Pseudomonas aeruginosa biofilms mediates phosphodiesterase activity, decreased cyclic Di-GMP levels, and enhanced dispersal. J Bacteriol 191(23):7333–7342. doi:10.1128/jb.00975-09

Bastari K, Arshath M, Ng Z, Chia J, Yow Z, Sana B, Tan M, Lim S, Loo S (2014) A controlled release of antibiotics from calcium phosphate-coated poly(lactic-co-glycolic acid) particles and their in vitro efficacy against Staphylococcus aureus biofilm. J Mater Sci Mater Med 25(3):747–757. doi:10.1007/s10856-013-5125-9

Bazaka K, Jacob MV, Crawford RJ, Ivanova EP (2011) Plasma-assisted surface modification of organic biopolymers to prevent bacterial attachment. Acta Biomater 7(5):2015–2028. http://dx.doi.org/10.1016/j.actbio.2010.12.024

Bazaka K, Jacob M, Crawford R, Ivanova E (2012) Efficient surface modification of biomaterial to prevent biofilm formation and the attachment of microorganisms. Appl Microbiol Biotechnol 95(2):299–311. doi:10.1007/s00253-012-4144-7

Beenken KE, Smith JK, Skinner RA, McLaren SG, Bellamy W, Gruenwald MJ, Spencer HJ, Jennings JA, Haggard WO, Smeltzer MS (2014) Chitosan coating to enhance the therapeutic efficacy of calcium sulfate-based antibiotic therapy in the treatment of chronic osteomyelitis. J Biomater Appl. doi:10.1177/0885328214535452

Bello A, Quinn M, Perry M, Milton D (2009) Characterization of occupational exposures to cleaning products used for common cleaning tasks-a pilot study of hospital cleaners. Environ Heal 8(1):11

Beloin C, Renard S, Ghigo J-M, Lebeaux D (2014) Novel approaches to combat bacterial biofilms. Curr Opin Pharmacol 18:61–68. http://dx.doi.org/10.1016/j.coph.2014.09.005

Bielecki TM, Gazdzik TS, Arendt J, Szczepanski T, Król W, Wielkoszynski T (2007) Antibacterial effect of autologous platelet gel enriched with growth factors and other active substances: an in vitro study. J Bone Joint Surg Br 89-B(3):417–420. doi:10.1302/0301-620x.89b3.18491

Billings N, Ramirez Millan M, Caldara M, Rusconi R, Tarasova Y, Stocker R, Ribbeck K (2013) The extracellular matrix component Psl provides fast-acting antibiotic defense in <italic>Pseudomonas aeruginosa</italic> iofilms. PLoS Pathog 9(8):e1003526. doi:10.1371/journal.ppat.1003526

Bower CK, Parker JE, Higgins AZ, Oest ME, Wilson JT, Valentine BA, Bothwell MK, McGuire J (2002) Protein antimicrobial barriers to bacterial adhesion: in vitro and in vivo evaluation of nisin-treated implantable materials. Colloids Surf B: Biointerfaces 25(1):81–90. http://dx.doi.org/10.1016/S0927-7765(01)00318-6

Braff MH, Zaiou M, Fierer J, Nizet V, Gallo RL (2005) Keratinocyte production of cathelicidin provides direct activity against bacterial skin pathogens. Infect Immun 73(10):6771–6781. doi:10.1128/iai.73.10.6771-6781.2005

Braydich-Stolle L, Hussain S, Schlager JJ, Hofmann M-C (2005) In vitro cytotoxicity of nanoparticles in mammalian germline stem cells. Toxicol Sci 88(2):412–419. doi:10.1093/toxsci/kfi256

Brogden KA (2005) Antimicrobial peptides: pore formers or metabolic inhibitors in bacteria? Nat Rev Microbiol 3(3):238–250. doi:10.1038/nrmicro1098

Brogden NK, Brogden KA (2011) Will new generations of modified antimicrobial peptides improve their potential as pharmaceuticals? Int J Antimicrob Agents 38(3):217–225. http://dx.doi.org/10.1016/j.ijantimicag.2011.05.004

Cashman J, Jackson J, Mugabe C, Gilchrist S, Ball K, Tredwell S, Burt H (2013) The use of tissue sealants to deliver antibiotics to an orthopaedic surgical site with a titanium implant. J Orthop Sci 18(1):165–174. doi:10.1007/s00776-012-0325-6

Cava F, de Pedro MA (2014) Peptidoglycan plasticity in bacteria: emerging variability of the murein sacculus and their associated biological functions. Curr Opin Microbiol 18:46–53. http://dx.doi.org/10.1016/j.mib.2014.01.004

Cegelski L, Pinkner JS, Hammer ND, Cusumano CK, Hung CS, Chorell E, Åberg V, Walker JN, Seed PC, Almqvist F, Chapman MR, Hultgren SJ (2009) Small-molecule inhibitors target

Escherichia coli amyloid biogenesis and biofilm formation. Nat Chem Biol 5(12):913–919. doi:10.1038/nchembio.242

Chandorkar Y, Bhaskar N, Madras G, Basu B (2015) Long term, sustained release of salicylic acid from crosslinked. Biodegradable polyesters induces reduced foreign body response in mice. Biomacromolecules. doi:10.1021/bm5017282

Chauhan A, Bernardin A, Mussard W, Kriegel I, Estève M, Ghigo J-M, Beloin C, Semetey V (2014) Preventing biofilm formation and associated occlusion by biomimetic glycocalyx-like polymer in central venous catheters. J Infect Dis. doi:10.1093/infdis/jiu249

Chen X, Su Y, Shen F, Wan Y (2011) Antifouling ultrafiltration membranes made from PAN-b-PEG copolymers: effect of copolymer composition and PEG chain length. J Membr Sci 384(1–2):44–51. http://dx.doi.org/10.1016/j.memsci.2011.09.002

Chen N-H, Chung C-J, Chiang C-C, Chen K-C, He J-L (2014) Antimicrobial copper-containing titanium nitride coatings Co-deposited by arc ion plating/magnetron sputtering for protective and decorative purposes. Surf Coat Technol 253:83–88. http://dx.doi.org/10.1016/j.surfcoat.2014.05.017

Cheng G, Xue H, Zhang Z, Chen S, Jiang S (2008) A switchable biocompatible polymer surface with self-sterilizing and nonfouling capabilities. Angew Chem Int Ed 47(46):8831–8834. doi:10.1002/anie.200803570

Cheng L, Weir MD, Xu HHK, Antonucci JM, Kraigsley AM, Lin NJ, Lin-Gibson S, Zhou X (2012) Antibacterial amorphous calcium phosphate nanocomposites with a quaternary ammonium dimethacrylate and silver nanoparticles. Dent Mater 28(5):561–572. http://dx.doi.org/10.1016/j.dental.2012.01.005

Chernousova S, Epple M (2013) Silver as antibacterial agent: ion, nanoparticle, and metal. Angew Chem Int Ed 52(6):1636–1653. doi:10.1002/anie.201205923

Chorley B, Ward W, Simmons SO, Vallanat B, Veronesi B (2014) The cellular and genomic response of rat dopaminergic neurons (N27) to coated nanosilver. NeuroToxicology 45:12–21. http://dx.doi.org/10.1016/j.neuro.2014.08.010

Christian WV, Oliver LD, Paustenbach DJ, Kreider ML, Finley BL (2014) Toxicology-based cancer causation analysis of CoCr-containing hip implants: a quantitative assessment of genotoxicity and tumorigenicity studies. J Appl Toxicol 34(9):939–967. doi:10.1002/jat.3039

Claessen D, Rozen DE, Kuipers OP, Sogaard-Andersen L, van Wezel GP (2014) Bacterial solutions to multicellularity: a tale of biofilms, filaments and fruiting bodies. Nat Rev Microbiol 12(2):115–124. doi:10.1038/nrmicro3178

Claridge JA, Banerjee A, Kelly KB, Leukhardt WH, Carter JW, Haridas M, Malangoni MA (2014) Bacterial species-specific hospital mortality rate for intra-abdominal infections. Surg Infect 15(3):194–199. doi:10.1089/sur.2011.039

Cleophas RTC, Riool M, Quarles Van Ufford HC, Zaat SAJ, Kruijtzer JAW, Liskamp RMJ (2014) Convenient preparation of bactericidal hydrogels by covalent attachment of stabilized antimicrobial peptides using thiol–ene click chemistry. ACS Macro Lett 3(5):477–480. doi:10.1021/mz5001465

Courvalin P (2006) Vancomycin resistance in gram-positive cocci. Clin Infect Dis 42(Supplement 1):S25–S34. doi:10.1086/491711

Cronholm P, Karlsson HL, Hedberg J, Lowe TA, Winnberg L, Elihn K, Wallinder IO, Möller L (2013) Intracellular uptake and toxicity of Ag and CuO nanoparticles: a comparison between nanoparticles and their corresponding metal ions. Small 9(7):970–982. doi:10.1002/smll.201201069

Duong HTT, Jung K, Kutty SK, Agustina S, Adnan NNM, Basuki JS, Kumar N, Davis TP, Barraud N, Boyer C (2014) Nanoparticle (star polymer) delivery of nitric oxide effectively negates Pseudomonas aeruginosa biofilm formation. Biomacromolecules 15(7):2583–2589. doi:10.1021/bm500422v

Dutot M, Liang H, Pauloin T, Brignole-Baudouin F, Baudouin C, Warnet J-M, Rat P (2008a) Effects of toxic cellular stresses and divalent cations on the human P2X7 cell death receptor. Mol Vis 14:889–897

Dutot M, Warnet J-M, Baudouin C, Rat P (2008b) Cytotoxicity of contact lens multipurpose solutions: role of oxidative stress, mitochondrial activity and P2X7 cell death receptor activation. Eur J Pharm Sci 33(2):138–145. http://dx.doi.org/10.1016/j.ejps.2007.10.006

Eckert R, Qi F, Yarbrough DK, He J, Anderson MH, Shi W (2006) Adding selectivity to antimicrobial peptides: rational design of a multidomain peptide against Pseudomonas spp. Antimicrob Agents Chemother 50(4):1480–1488. doi:10.1128/aac.50.4.1480-1488.2006

Etayash H, Norman L, Thundat T, Stiles M, Kaur K (2013) Surface-conjugated antimicrobial peptide leucocin a displays high binding to pathogenic gram-positive bacteria. ACS Appl Mater Interfaces 6(2):1131–1138. doi:10.1021/am404729c

Faber M, Leysen J, Bridts C, Sabato V, De Clerck LS, Ebo DG (2012) Allergy to chlorhexidine: beware of the central venous catheter. Acta Anaesthesiol Belg 63(4):191–194

Fadeeva E, Truong VK, Stiesch M, Chichkov BN, Crawford RJ, Wang J, Ivanova EP (2011) Bacterial retention on superhydrophobic titanium surfaces fabricated by femtosecond laser ablation. Langmuir 27(6):3012–3019. doi:10.1021/la104607g

Fan JH, Hung WI, Li WT, Yeh JM (2009) Biocompatibility study of gold nanoparticles to human cells. In: Lim C, Goh JH (eds) 13th international conference on biomedical engineering, vol 23. IFMBE proceedings. Springer, Berlin, pp 870–873. doi:10.1007/978-3-540-92841-6_214

Fernandez-Lopez S, Kim H-S, Choi EC, Delgado M, Granja JR, Khasanov A, Kraehenbuehl K, Long G, Weinberger DA, Wilcoxen KM, Ghadiri MR (2001) Antibacterial agents based on the cyclic d, l-[alpha]-peptide architecture. Nature 412(6845):452–455

Fitzgerald-Hughes D, Devocelle M, Humphreys H (2012) Beyond conventional antibiotics for the future treatment of methicillin-resistant Staphylococcus aureus infections: two novel alternatives. FEMS Immunol Med Microbiol 65(3):399–412. doi:10.1111/j.1574-695X.2012.00954.x

Forbes S, McBain AJ, Felton-Smith S, Jowitt TA, Birchenough HL, Dobson CB (2013) Comparative surface antimicrobial properties of synthetic biocides and novel human apolipoprotein E derived antimicrobial peptides. Biomaterials 34(22):5453–5464. http://dx.doi.org/10.1016/j.biomaterials.2013.03.087

Francolini I, Donelli G (2010) Prevention and control of biofilm-based medical-device-related infections. FEMS Immunol Med Microbiol 59(3):227–238. doi:10.1111/j.1574-695X.2010.00665.x

Francolini I, Donelli G, Crisante F, Taresco V, Piozzi A (2015) Antimicrobial polymers for anti-biofilm medical devices: state-of-art and perspectives. In: Donelli G (ed) Biofilm-based healthcare-associated infections. Advances in experimental medicine and biology, vol 831. Springer International Publishing, Switzerland, pp 93–117. doi:10.1007/978-3-319-09782-4_7

Friedlander RS, Vlamakis H, Kim P, Khan M, Kolter R, Aizenberg J (2013) Bacterial flagella explore microscale hummocks and hollows to increase adhesion. Proc Natl Acad Sci 110(14):5624–5629. doi:10.1073/pnas.1219662110

Fu C, Liu T, Li L, Liu H, Liang Q, Meng X (2015) Effects of graphene oxide on the development of offspring mice in lactation period. Biomaterials 40:23–31. http://dx.doi.org/10.1016/j.biomaterials.2014.11.014

Gajski G, Jelčić Ž, Oreščanin V, Gerić M, Kollar R, Garaj-Vrhovac V (2014) Physico-chemical characterization and the in vitro genotoxicity of medical implants metal alloy (TiAlV and CoCrMo) and polyethylene particles in human lymphocytes. Biochim Biophys Acta Gen Subj 1840(1):565–576. http://dx.doi.org/10.1016/j.bbagen.2013.10.015

Gan X, Liu T, Zhong J, Liu X, Li G (2004) Effect of silver nanoparticles on the electron transfer reactivity and the catalytic activity of myoglobin. ChemBioChem 5(12):1686–1691. doi:10.1002/cbic.200400080

Gao P, Nie X, Zou M, Shi Y, Cheng G (2011) Recent advances in materials for extended-release antibiotic delivery system. J Antibiot 64(9):625–634

Gening ML, Titov DV, Cecioni S, Audfray A, Gerbst AG, Tsvetkov YE, Krylov VB, Imberty A, Nifantiev NE, Vidal S (2013) Synthesis of multivalent carbohydrate-centered glycoclusters as nanomolar ligands of the bacterial lectin LecA from Pseudomonas aeruginosa. Chem Eur J 19(28):9272–9285. doi:10.1002/chem.201300135

Gérard C, Bordeleau L-J, Barralet J, Doillon CJ (2010) The stimulation of angiogenesis and collagen deposition by copper. Biomaterials 31(5):824–831. http://dx.doi.org/10.1016/j.biomaterials.2009.10.009

Gerdes K, Maisonneuve E (2012) Bacterial persistence and toxin-antitoxin loci. Annu Rev Microbiol 66(1):103–123. doi:10.1146/annurev-micro-092611-150159

Germain E, Castro-Roa D, Zenkin N, Gerdes K (2013) Molecular mechanism of bacterial persistence by HipA. Mol Cell 52(2):248–254. doi:10.1016/j.molcel.2013.08.045

Gnanadhas DP, Ben Thomas M, Thomas R, Raichur AM, Chakravortty D (2013) Interaction of silver nanoparticles with serum proteins affects their antimicrobial activity in vivo. Antimicrob Agents Chemother 57(10):4945–4955. doi:10.1128/aac.00152-13

Grainger DW, van der Mei HC, Jutte PC, van den Dungen JJAM, Schultz MJ, van der Laan BFAM, Zaat SAJ, Busscher HJ (2013) Critical factors in the translation of improved antimicrobial strategies for medical implants and devices. Biomaterials 34(37):9237–9243. http://dx.doi.org/10.1016/j.biomaterials.2013.08.043

Guiton PS, Cusumano CK, Kline KA, Dodson KW, Han Z, Janetka JW, Henderson JP, Caparon MG, Hultgren SJ (2012) Combinatorial small-molecule therapy prevents uropathogenic Escherichia coli catheter-associated urinary tract infections in mice. Antimicrob Agents Chemother 56(9):4738–4745. doi:10.1128/aac.00447-12

Guleri A, Kumar A, Morgan RJM, Hartley M, Roberts DH (2012) Anaphylaxis to chlorhexidine-coated central venous catheters: a case series and review of the literature. Surg Infect 13(3):171–174. doi:10.1089/sur.2011.011

Hale JDF, Hancock REW (2007) Alternative mechanisms of action of cationic antimicrobial peptides on bacteria. Expert Rev Anti-Infect Ther 5(6):951–959. doi:10.1586/14787210.5.6.951

Hamuro Y, Schneider JP, Degrado WF (1999) De novo design of antibacterial β-peptides. J Am Chem Soc 121(51):12200–12201. doi:10.1021/ja992728p

Hancock REW, Sahl H-G (2006) Antimicrobial and host-defense peptides as new anti-infective therapeutic strategies. Nat Biotech 24(12):1551–1557

Hegstad K, Langsrud S, Lunestad BT, Scheie AA, Sunde M, Yazdankhah SP (2010) Does the wide use of quaternary ammonium compounds enhance the selection and spread of antimicrobial resistance and thus threaten our health? Microb Drug Resist 16:91+

Hennessey DB, Burke JP, Ni-Dhonochu T, Shields C, Winter DC, Mealy K (2010) Preoperative hypoalbuminemia is an independent risk factor for the development of surgical site infection following gastrointestinal surgery: a multi-institutional study. Ann Surg 252(2):325–329 310.1097/SLA.1090b1013e3181e9819a

Hogan S, Stevens NT, Humphreys H, O'Gara JP, O'Neill E (2015) Current and future approaches to the prevention and treatment of staphylococcal medical device-related infections. Curr Pharm Des 21(1):100–113

Holden DW (2015) Persisters unmasked. Science 347(6217):30–32. doi:10.1126/science.1262033

Hook AL, Chang C-Y, Yang J, Luckett J, Cockayne A, Atkinson S, Mei Y, Bayston R, Irvine DJ, Langer R, Anderson DG, Williams P, Davies MC, Alexander MR (2012) Combinatorial discovery of polymers resistant to bacterial attachment. Nat Biotechnol 30(9):868–875. doi:10.1038/nbt.2316. http://www.nature.com/nbt/journal/v30/n9/abs/nbt.2316.html#supplementary-information

Hsin Y-H, Chen C-F, Huang S, Shih T-S, Lai P-S, Chueh PJ (2008) The apoptotic effect of nanosilver is mediated by a ROS- and JNK-dependent mechanism involving the mitochondrial pathway in NIH3T3 cells. Toxicol Lett 179(3):130–139. http://dx.doi.org/10.1016/j.toxlet.2008.04.015

Imura Y, Choda N, Matsuzaki K (2008) Magainin 2 in action: distinct modes of membrane permeabilization in living bacterial and mammalian cells. Biophys J 95(12):5757–5765. http://dx.doi.org/10.1529/biophysj.108.133488

Ivanova EP, Hasan J, Webb HK, Truong VK, Watson GS, Watson JA, Baulin VA, Pogodin S, Wang JY, Tobin MJ, Löbbe C, Crawford RJ (2012) Natural bactericidal surfaces: mechanical rupture of Pseudomonas aeruginosa cells by cicada wings. Small 8(16):2489–2494. doi:10.1002/smll.201200528

Ivanova EP, Hasan J, Webb HK, Gervinskas G, Juodkazis S, Truong VK, Wu AHF, Lamb RN, Baulin VA, Watson GS, Watson JA, Mainwaring DE, Crawford RJ (2013) Bactericidal activity of black silicon. Nat Commun 4. doi:10.1038/ncomms3838

Jacobs T, Morent R, De Geyter N, Dubruel P, Leys C (2012) Plasma surface modification of biomedical polymers: influence on cell-material interaction. Plasma Chem Plasma Process 32(5):1039–1073. doi:10.1007/s11090-012-9394-8

Jadalannagari S, Deshmukh K, Ramanan S, Kowshik M (2014) Antimicrobial activity of hemocompatible silver doped hydroxyapatite nanoparticles synthesized by modified sol–gel technique. Appl Nanosci 4(2):133–141. doi:10.1007/s13204-013-0197-x

Jamal MA, Rosenblatt JS, Hachem RY, Ying J, Pravinkumar E, Nates JL, Chaftari A-MP, Raad II (2014) Prevention of biofilm colonization by Gram-negative bacteria on minocycline-rifampin-impregnated catheters sequentially coated with chlorhexidine. Antimicrob Agents Chemother 58(2):1179–1182. doi:10.1128/aac.01959-13

Jayathillake A, Mason DF, Broome K (2003) Allergy to chlorhexidine gluconate in urethral gel: report of four cases and review of the literature. Urology 61(4):837. doi:10.1016/S0090-4295(02)02432-9

Jo YK, Seo JH, Choi B-H, Kim BJ, Shin HH, Hwang BH, Cha HJ (2014) Surface-independent antibacterial coating using silver nanoparticle-generating engineered mussel glue. ACS Appl Mater Interfaces 6(22):20242–20253. doi:10.1021/am505784k

Johnson JR, Johnston B, Kuskowski MA (2012) In vitro comparison of nitrofurazone- and silver alloy-coated foley catheters for contact-dependent and diffusible inhibition of urinary tract infection-associated microorganisms. Antimicrob Agents Chemother 56(9):4969–4972. doi:10.1128/aac.00733-12

Jun E-A, Lim K-M, Kim K, Bae O-N, Noh J-Y, Chung K-H, Chung J-H (2011) Silver nanoparticles enhance thrombus formation through increased platelet aggregation and procoagulant activity. Nanotoxicology 5(2):157–167. doi:10.3109/17435390.2010.506250

Justice SS, Harrison A, Becknell B, Mason KM (2014) Bacterial differentiation, development, and disease: mechanisms for survival. FEMS Microbiol Lett 360(1):1–8. doi:10.1111/1574-6968.12602

Kalaivani S, Singh RK, Ganesan V, Kannan S (2014) Effect of copper (Cu2+) inclusion on the bioactivity and antibacterial behavior of calcium silicate coatings on titanium metal. J Mater Chem B 2(7):846–858. doi:10.1039/C3TB21522A

Karahalil B, Kadioglu E, Tuzuner-Oncul AM, Cimen E, Emerce E, Kisnisci RS (2014) Micronucleus assay assessment of possible genotoxic effects in patients treated with titanium alloy endosseous implants or miniplates. Mutat Res/Genet Toxicol Environ Mutagen 760:70–72. http://dx.doi.org/10.1016/j.mrgentox.2013.10.005

Karlsson HL, Cronholm P, Hedberg Y, Tornberg M, De Battice L, Svedhem S, Wallinder IO (2013) Cell membrane damage and protein interaction induced by copper containing nanoparticles—Importance of the metal release process. Toxicology 313(1):59–69. http://dx.doi.org/10.1016/j.tox.2013.07.012

Karlsson J, Sundell G, Thuvander M, Andersson M (2014) Atomically resolved tissue integration. Nano Lett 14(8):4220–4223. doi:10.1021/nl501564f

Kawata K, Osawa M, Okabe S (2009) In vitro toxicity of silver nanoparticles at noncytotoxic doses to HepG2 human hepatoma cells. Environ Sci Technol 43(15):6046–6051. doi:10.1021/es900754q

Khan W, Challa V, Pawar R, Nyska M, Brin Y, Domb A (2014) Antibiotics delivery for treating bone infections. In: Domb AJ, Khan W (eds) Focal controlled drug delivery. Advances in delivery science and technology. Springer, New York, pp 459–472. doi:10.1007/978-1-4614-9434-8_21

Khoo A, Oziemski P (2011) Chlorhexidine impregnated central venous catheter inducing an anaphylactic shock in the intensive care unit. Heart Lung Circ 20(10):669–670. http://dx.doi.org/10.1016/j.hlc.2010.10.001

Kilic E, Er N, Alkan A, Ferahbas A (2011) Accidental benzalkonium chloride (zephiran) injection. Oral Surg Oral Med Oral Pathol Oral Radiol Endod 112(6):e103–e105. doi:10.1016/j.tripleo.2011.06.004

Kim S, Choi JE, Choi J, Chung K-H, Park K, Yi J, Ryu D-Y (2009) Oxidative stress-dependent toxicity of silver nanoparticles in human hepatoma cells. Toxicol in Vitro 23(6):1076–1084. http://dx.doi.org/10.1016/j.tiv.2009.06.001

Knetsch MLW, Koole LH (2011) New strategies in the development of antimicrobial coatings: the example of increasing usage of silver and silver nanoparticles. Polymers 3(1):340–366

Kottur J, Sharma A, Gore Kiran R, Narayanan N, Samanta B, Pradeepkumar PI, Nair Deepak T (2015) Unique structural features in DNA polymerase IV enable efficient bypass of the N2 adduct induced by the nitrofurazone antibiotic. Structure 23(1):56–67. http://dx.doi.org/10.1016/j.str.2014.10.019

Lang NP, Salvi GE, Huynh-Ba G, Ivanovski S, Donos N, Bosshardt DD (2011) Early osseointegration to hydrophilic and hydrophobic implant surfaces in humans. Clin Oral Implants Res 22(4):349–356. doi:10.1111/j.1600-0501.2011.02172.x

Laroux FS, Pavlick KP, Hines IN, Kawachi S, Harada H, Bharwani S, Hoffman JM, Grisham MB (2001) Role of nitric oxide in inflammation. Acta Physiol Scand 173(1):113–118. doi:10.1046/j.1365-201X.2001.00891.x

Lenaeus MJ, Burdette D, Wagner T, Focia PJ, Gross A (2014) Structures of KcsA in complex with symmetrical quaternary ammonium compounds reveal a hydrophobic binding site. Biochemistry 53(32):5365–5373. doi:10.1021/bi500525s

Li P, Poon YF, Li W, Zhu H-Y, Yeap SH, Cao Y, Qi X, Zhou C, Lamrani M, Beuerman RW, Kang E-T, Mu Y, Li CM, Chang MW, Jan Leong SS, Chan-Park MB (2011) A polycationic antimicrobial and biocompatible hydrogel with microbe membrane suctioning ability. Nat Mater 10(2):149–156. http://www.nature.com/nmat/journal/v10/n2/abs/nmat2915.html#supplementary-information

Li X, Wang L, Fan Y, Feng Q, Cui F-z (2012) Biocompatibility and toxicity of nanoparticles and nanotubes. J Nanomater 2012:19. doi:10.1155/2012/548389

Li Y, Heine S, Entian M, Sauer K, Frankenberg-Dinkel N (2013) NO-induced biofilm dispersion in Pseudomonas aeruginosa is mediated by an MHYT domain-coupled phosphodiesterase. J Bacteriol 195(16):3531–3542. doi:10.1128/jb.01156-12

Lih E, Oh SH, Joung YK, Lee JH, Han DK (2015) Polymers for cell/tissue anti-adhesion. Progress in Polymer Science. http://dx.doi.org/10.1016/j.progpolymsci.2014.10.004

Ling LL, Schneider T, Peoples AJ, Spoering AL, Engels I, Conlon BP, Mueller A, Schaberle TF, Hughes DE, Epstein S, Jones M, Lazarides L, Steadman VA, Cohen DR, Felix CR, Fetterman KA, Millett WP, Nitti AG, Zullo AM, Chen C, Lewis K (2015) A new antibiotic kills pathogens without detectable resistance. Nat Adv Online Publ. doi:10.1038/nature14098. http://www.nature.com/nature/journal/vaop/ncurrent/abs/nature14098.html#supplementary-information

Lisanti M, Piolanti N, Tagliaferri E, Andreani L, Parchi P, Menichetti F (2015) The fight against the slime: can we ever win? In: Baldini A, Caldora P (eds) Perioperative medical management for total joint arthroplasty. Springer International Publishing, Switzerland, pp 253–267. doi:10.1007/978-3-319-07203-6_21

Liu S, Zeng TH, Hofmann M, Burcombe E, Wei J, Jiang R, Kong J, Chen Y (2011) Antibacterial activity of graphite, graphite oxide, graphene oxide, and reduced graphene oxide: membrane and oxidative stress. ACS Nano 5(9):6971–6980. doi:10.1021/nn202451x

Liu H, Zhang D, Shen F, Zhang G, Song S (2012a) Hemocompatibility and anti-endothelialization of copper–titanium coating for vena cava filters. Surf Coat Technol 206(16):3501–3507. http://dx.doi.org/10.1016/j.surfcoat.2012.02.025

Liu W-d, Zhang Y-h, Fang L-f, Zhu B-k, Zhu L-p (2012b) Antifouling properties of poly(vinyl chloride) membranes modified by amphiphilic copolymers P(MMA-b-MAA). Chin J Polym Sci 30(4):568–577. doi:10.1007/s10118-012-1153-z

Liu X, Xu Y, Wu Z, Chen H (2013a) Poly(N-vinylpyrrolidone)-modified surfaces for biomedical applications. Macromol Biosci 13(2):147–154. doi:10.1002/mabi.201200269

Liu Y, Yang Y, Wang C, Zhao X (2013b) Stimuli-responsive self-assembling peptides made from antibacterial peptides. Nanoscale 5(14):6413–6421. doi:10.1039/c3nr00225j

Liu X, Huang R, Su R, Qi W, Wang L, He Z (2014) Grafting hyaluronic acid onto gold surface to achieve low protein fouling in surface plasmon resonance biosensors. ACS Appl Mater Interfaces 6(15):13034–13042. doi:10.1021/am502921z

Lo AWH, Van de Water K, Gane PJ, Chan AWE, Steadman D, Stevens K, Selwood DL, Waksman G, Remaut H (2013) Suppression of type 1 pilus assembly in uropathogenic Escherichia coli by chemical inhibition of subunit polymerization. J Antimicrob Chemother. doi:10.1093/jac/dkt467

Ma J, Zhang J, Xiong Z, Yong Y, Zhao XS (2011) Preparation, characterization and antibacterial properties of silver-modified graphene oxide. J Mater Chem 21(10):3350–3352. doi:10.1039/C0JM02806A

Maeda H, Uehara A, Saito T, Mayanagi H, Nagaoka I, Takada H (2007) An antibacterial protein CAP18/LL-37 enhanced production of hepatocyte growth factor in human gingival fibroblast cultures. In: Watanabe M, Okuno O, Sasaki K, Takahashi N, Suzuki O, Takada H (eds) Interface oral health science 2007. Springer, Tokyo, Japan, pp 283–284. doi:10.1007/978-4-431-76690-2_55

Maisonneuve E, Shakespeare LJ, Jørgensen MG, Gerdes K (2011) Bacterial persistence by RNA endonucleases. Proc Natl Acad Sci 108(32):13206–13211. doi:10.1073/pnas.1100186108

Martín VI, de la Haba RR, Ventosa A, Congiu E, Ortega-Calvo JJ, Moyá ML (2014) Colloidal and biological properties of cationic single-chain and dimeric surfactants. Colloids Surf B: Biointerfaces 114:247–254. http://dx.doi.org/10.1016/j.colsurfb.2013.10.017

Mazzotti F, Beuttler J, Zeller R, Fink U, Schindler S, Wendel A, Hartung T, von Aulock S (2007) In vitro pyrogen test—a new test method for solid medical devices. J Biomed Mater Res A 80A(2):276–282. doi:10.1002/jbm.a.30922

Miura N, Shinohara Y (2009) Cytotoxic effect and apoptosis induction by silver nanoparticles in HeLa cells. Biochem Biophys Res Commun 390(3):733–737. http://dx.doi.org/10.1016/j.bbrc.2009.10.039

Moalli P, Brown B, Reitman MF, Nager C (2014) Polypropylene mesh: evidence for lack of carcinogenicity. Int Urogynecol J 25(5):573–576. doi:10.1007/s00192-014-2343-8

Morones-Ramirez JR, Winkler JA, Spina CS, Collins JJ (2013) Silver enhances antibiotic activity against Gram-negative bacteria. Sci Transl Med 5(190):190ra181. doi:10.1126/scitranslmed.3006276

Mouriño V, Cattalini JP, Boccaccini AR (2012) Metallic ions as therapeutic agents in tissue engineering scaffolds: an overview of their biological applications and strategies for new developments. J R Soc Interface 9(68):401–419

Muffly TM, Tizzano AP, Walters MD (2011) The history and evolution of sutures in pelvic surgery. J R Soc Med 104(3):107–112. doi:10.1258/jrsm.2010.100243

Muszanska AK, Rochford ETJ, Gruszka A, Bastian AA, Busscher HJ, Norde W, van der Mei HC, Herrmann A (2014) Antiadhesive polymer brush coating functionalized with antimicrobial and RGD peptides to reduce biofilm formation and enhance tissue integration. Biomacromolecules 15(6):2019–2026. doi:10.1021/bm500168s

Nait Chabane Y, Mlouka M, Alexandre S, Nicol M, Marti S, Pestel-Caron M, Vila J, Jouenne T, De E (2014) Virstatin inhibits biofilm formation and motility of Acinetobacter baumannii. BMC Microbiol 14(1):62

Neumayer L, Hosokawa P, Itani K, El-Tamer M, Henderson WG, Khuri SF (2007) Multivariable predictors of postoperative surgical site infection after general and vascular surgery: results from the patient safety in surgery study. J Am Coll Surg 204(6):1178–1187. http://dx.doi.org/10.1016/j.jamcollsurg.2007.03.022

Ng VWL, Tan JPK, Leong J, Voo ZX, Hedrick JL, Yang YY (2014) Antimicrobial polycarbonates: investigating the impact of nitrogen-containing heterocycles as quaternizing agents. Macromolecules 47(4):1285–1291. doi:10.1021/ma402641p

Nikitushkin V, Demina G, Shleeva M, Kaprelyants A (2013) Peptidoglycan fragments stimulate resuscitation of "non-culturable" mycobacteria. Antonie Van Leeuwenhoek 103(1):37–46. doi:10.1007/s10482-012-9784-1

Nizet V, Ohtake T, Lauth X, Trowbridge J, Rudisill J, Dorschner RA, Pestonjamasp V, Piraino J, Huttner K, Gallo RL (2001) Innate antimicrobial peptide protects the skin from invasive bacterial infection. Nature 414(6862):454–457

O'Connell HA, Kottkamp GS, Eppelbaum JL, Stubblefield BA, Gilbert SE, Gilbert ES (2006) Influences of biofilm structure and antibiotic resistance mechanisms on indirect pathogenicity in a model polymicrobial biofilm. Appl Environ Microbiol 72(7):5013–5019. doi:10.1128/aem.02474-05

O'Toole JE, Eichholz KM, Fessler RG (2009) Surgical site infection rates after minimally invasive spinal surgery. J Neurosurg Spine 11(4):471–476. doi:10.3171/2009.5.SPINE08633

Olsen MA, Nepple JJ, Riew KD, Lenke LG, Bridwell KH, Mayfield J, Fraser VJ (2008) Risk factors for surgical site infection following orthopaedic spinal operations. J Bone Joint Surg 90(1):62–69

Opstrup MS, Malling HJ, Krøigaard M, Mosbech H, Skov PS, Poulsen LK, Garvey LH (2014) Standardized testing with chlorhexidine in perioperative allergy – a large single-centre evaluation. Allergy 69(10):1390–1396. doi:10.1111/all.12466

Ordikhani F, Tamjid E, Simchi A (2014) Characterization and antibacterial performance of electrodeposited chitosan–vancomycin composite coatings for prevention of implant-associated infections. Mater Sci Eng C 41:240–248. http://dx.doi.org/10.1016/j.msec.2014.04.036

Oren Z, Shai Y (2000) Cyclization of a cytolytic amphipathic α-helical peptide and its diastereomer: effect on structure, interaction with model membranes, and biological function†. Biochemistry 39(20):6103–6114. doi:10.1021/bi992408i

Owens CD, Stoessel K (2008) Surgical site infections: epidemiology, microbiology and prevention. J Hosp Infect 70(Supplement 2):3–10. http://dx.doi.org/10.1016/S0195-6701(08)60017-1

Panyutich AV, Voitenok NN, Lehrer RI, Ganz T (1991) An enzyme immunoassay for human defensins. J Immunol Methods 141(2):149–155. http://dx.doi.org/10.1016/0022-1759(91)90141-2

Park E-J, Yi J, Kim Y, Choi K, Park K (2010) Silver nanoparticles induce cytotoxicity by a Trojan-horse type mechanism. Toxicol in Vitro 24(3):872–878. http://dx.doi.org/10.1016/j.tiv.2009.12.001

Patra P, Klumpp S (2013) Population dynamics of bacterial persistence. PLoS ONE 8(5):e62814. doi:10.1371/journal.pone.0062814

Perez E, Williams M, Jacob JT, Reyes MD, Chernetsky Tejedor S, Steinberg JP, Rowe L, Ganakammal SR, Changayil S, Weil MR, Donlan RM (2014) Microbial biofilms on needleless connectors for central venous catheters: comparison of standard and silver-coated devices collected from patients in an acute care hospital. J Clin Microbiol 52(3):823–831. doi:10.1128/jcm.02220-13

Phillips J, O'Grady H, Baker E (2014) Prevention of surgical site infections. Surgery (Oxford) 32(9):468–471. http://dx.doi.org/10.1016/j.mpsur.2014.06.011

Pittet B, Montandon D, Pittet D (2005) Infection in breast implants. Lancet Infect Dis 5(2):94–106. http://dx.doi.org/10.1016/S1473-3099(05)01281-8

Podila R, Moore T, Alexis F, Rao AM (2013) Graphene coatings for enhanced hemo-compatibility of nitinol stents. RSC Adv 3(6):1660–1665. doi:10.1039/C2RA23073A

Porter EA, Wang X, Lee H-S, Weisblum B, Gellman SH (2000) Antibiotics: non-haemolytic [beta]-amino-acid oligomers. Nature 404(6778):565–565

Powers J-PS, Hancock REW (2003) The relationship between peptide structure and antibacterial activity. Peptides 24(11):1681–1691. http://dx.doi.org/10.1016/j.peptides.2003.08.023

Rai M, Yadav A, Gade A (2009) Silver nanoparticles as a new generation of antimicrobials. Biotechnol Adv 27(1):76–83. http://dx.doi.org/10.1016/j.biotechadv.2008.09.002
Rapsch K, Bier FF, Tadros M, von Nickisch-Rosenegk M (2013) Identification of antimicrobial peptides and immobilization strategy suitable for a covalent surface coating with biocompatible properties. Bioconjug Chem 25(2):308–319. doi:10.1021/bc4004469
Rathbone CR, Cross JD, Brown KV, Murray CK, Wenke JC (2011) Effect of various concentrations of antibiotics on osteogenic cell viability and activity. J Orthop Res 29(7):1070–1074. doi:10.1002/jor.21343
Ravikumar M, Hageman DJ, Tomaszewski WH, Chandra GM, Skousen JL, Capadona JR (2014) The effect of residual endotoxin contamination on the neuroinflammatory response to sterilized intracortical microelectrodes. J Mater Chem B 2(17):2517–2529. doi:10.1039/C3TB21453B
Ridgeway S, Wilson J, Charlet A, Kafatos G, Pearson A, Coello R (2005) Infection of the surgical site after arthroplasty of the hip. J Bone Joint Surg Br Vol 87-B(6):844–850. doi:10.1302/0301-620x.87b6.15121
Rosenthal VD (2008) Device-associated nosocomial infections in limited-resources countries: findings of the International Nosocomial Infection Control Consortium (INICC). Am J Infect Control 36(10):S171.e177–S171.e112. http://dx.doi.org/10.1016/j.ajic.2008.10.009
Ryder VJ, Chopra I, O'Neill AJ (2012) Increased mutability of Staphylococci in biofilms as a consequence of oxidative stress. PLoS ONE 7(10):e47695. doi:10.1371/journal.pone.0047695
Sahl H-G, Pag U, Bonness S, Wagner S, Antcheva N, Tossi A (2005) Mammalian defensins: structures and mechanism of antibiotic activity. J Leukoc Biol 77(4):466–475. doi:10.1189/jlb.0804452
Salick DA, Kretsinger JK, Pochan DJ, Schneider JP (2007) Inherent antibacterial activity of a peptide-based β-hairpin hydrogel. J Am Chem Soc 129(47):14793–14799. doi:10.1021/ja076300z
Salick DA, Pochan DJ, Schneider JP (2009) Design of an injectable β-hairpin peptide hydrogel that kills methicillin-resistant Staphylococcus aureus. Adv Mater 21(41):4120–4123. doi:10.1002/adma.200900189
Salwiczek M, Qu Y, Gardiner J, Strugnell RA, Lithgow T, Mclean KM, Thissen H (2014) Emerging rules for effective antimicrobial coatings. Trends Biotechnol 32(2):82–90. http://dx.doi.org/10.1016/j.tibtech.2013.09.008
Sambanthamoorthy K, Sloup RE, Parashar V, Smith JM, Kim EE, Semmelhack MF, Neiditch MB, Waters CM (2012) Identification of small molecules that antagonize diguanylate cyclase enzymes to inhibit biofilm formation. Antimicrob Agents Chemother 56(10):5202–5211. doi:10.1128/aac.01396-12
Sambanthamoorthy K, Luo C, Pattabiraman N, Feng X, Koestler B, Waters CM, Palys TJ (2013) Identification of small molecules inhibiting diguanylate cyclases to control bacterial biofilm development. Biofouling 30(1):17–28. doi:10.1080/08927014.2013.832224
Savage VJ, Chopra I, O'Neill AJ (2013) Staphylococcus aureus biofilms promote horizontal transfer of antibiotic resistance. Antimicrob Agents Chemother 57(4):1968–1970. doi:10.1128/aac.02008-12
Shalev T, Gopin A, Bauer M, Stark RW, Rahimipour S (2012) Non-leaching antimicrobial surfaces through polydopamine bio-inspired coating of quaternary ammonium salts or an ultrashort antimicrobial lipopeptide. J Mater Chem 22(5):2026–2032. doi:10.1039/C1JM13994K
Shamir ER, Warthan M, Brown SP, Nataro JP, Guerrant RL, Hoffman PS (2010) Nitazoxanide inhibits biofilm production and hemagglutination by enteroaggregative Escherichia coli strains by blocking assembly of AafA fimbriae. Antimicrob Agents Chemother 54(4):1526–1533. doi:10.1128/aac.01279-09
Sharaf B, Jandali-Rifai M, Susarla SM, Dodson TB (2011) Do perioperative antibiotics decrease implant failure? J Oral Maxillofac Surg 69(9):2345–2350. http://dx.doi.org/10.1016/j.joms.2011.02.095
Sharma JN, Al-Omran A, Parvathy SS (2007) Role of nitric oxide in inflammatory diseases. Inflammopharmacology 15(6):252–259. doi:10.1007/s10787-007-0013-x

Shetye GS, Singh N, Jia C, Nguyen CDK, Wang G, Luk Y-Y (2014) Specific maltose derivatives modulate the swarming motility of nonswarming mutant and inhibit bacterial adhesion and biofilm formation by Pseudomonas aeruginosa. ChemBioChem 15(10):1514–1523. doi:10.1002/cbic.201402093

Shimazaki K-i, Nam MS, Harakawa S, Tanaka T, Omata Y, Saito A, Kumura H, Mikawa K, Igarashi I, Suzuki N (1996) Monoclonal antibody against bovine lactoferricin(R) and its epitopic site. J Vet Med Sci 58(12):1227–1229. doi:10.1292/jvms.58.12_1227

Shimizu M, Shigeri Y, Tatsu Y, Yoshikawa S, Yumoto N (1998) Enhancement of antimicrobial activity of neuropeptide Y by N-terminal truncation. Antimicrob Agents Chemother 42(10):2745–2746

Shrivastava S, Bera T, Singh SK, Singh G, Ramachandrarao P, Dash D (2009) Characterization of antiplatelet properties of silver nanoparticles. ACS Nano 3(6):1357–1364. doi:10.1021/nn900277t

Silvestri DL, Mcenery-Stonelake M (2013) Chlorhexidine: uses and adverse reactions. Dermatitis 24(3):112–118 110.1097/DER.1090b1013e3182905561

Singh PK, Schaefer AL, Parsek MR, Moninger TO, Welsh MJ, Greenberg EP (2000) Quorum-sensing signals indicate that cystic fibrosis lungs are infected with bacterial biofilms. Nature 407(6805):762–764

Skousen JL, Bridge MJ, Tresco PA (2015) A strategy to passively reduce neuroinflammation surrounding devices implanted chronically in brain tissue by manipulating device surface permeability. Biomaterials 36:33–43. http://dx.doi.org/10.1016/j.biomaterials.2014.08.039

Smock KJ, Schmidt RL, Hadlock G, Stoddard G, Grainger DW, Munger MA (2014) Assessment of orally dosed commercial silver nanoparticles on human ex vivo platelet aggregation. Nanotoxicology 8(3):328–333. doi:10.3109/17435390.2013.788749

Solano C, Echeverz M, Lasa I (2014) Biofilm dispersion and quorum sensing. Curr Opin Microbiol 18:96–104. http://dx.doi.org/10.1016/j.mib.2014.02.008

Sørensen O, Cowland JB, Askaa J, Borregaard N (1997) An ELISA for hCAP-18, the cathelicidin present in human neutrophils and plasma. J Immunol Methods 206(1–2):53–59. http://dx.doi.org/10.1016/S0022-1759(97)00084-7

Sumner DR, Ross R, Purdue E (2014) Are there biological markers for wear or corrosion? A systematic review. Clin Orthop Relat Res 472(12):3728–3739. doi:10.1007/s11999-014-3580-3

Świercz R, Hałatek T, Stetkiewicz J, Wąsowicz W, Kur B, Grzelińska Z, Majcherek W (2013) Toxic effect in the lungs of rats after inhalation exposure to benzalkonium chloride. IJOMEH 26(4):647–656. doi:10.2478/s13382-013-0137-8

Taheri S, Cavallaro A, Christo SN, Smith LE, Majewski P, Barton M, Hayball JD, Vasilev K (2014) Substrate independent silver nanoparticle based antibacterial coatings. Biomaterials 35(16):4601–4609. http://dx.doi.org/10.1016/j.biomaterials.2014.02.033

Takeshima K, Chikushi A, Lee K-K, Yonehara S, Matsuzaki K (2003) Translocation of analogues of the antimicrobial peptides magainin and buforin across human cell membranes. J Biol Chem 278(2):1310–1315. doi:10.1074/jbc.M208762200

Tamam E, Turkyilmaz I (2014) Effects of pH and elevated glucose levels on the electrochemical behavior of dental implants. J Oral Implantol 40(2):153–159. doi:10.1563/AAID-JOI-D-11-00083

Taylor A (1985) Therapeutic uses of trace elements. Clin Endocrinol Metab 14(3):703–724. http://dx.doi.org/10.1016/S0300-595X(85)80013-X

Taylor U, Barchanski A, Garrels W, Klein S, Kues W, Barcikowski S, Rath D (2012) Toxicity of gold nanoparticles on somatic and reproductive cells. In: Zahavy E, Ordentlich A, Yitzhaki S, Shafferman A (eds) Nano-biotechnology for biomedical and diagnostic research. Advances in experimental medicine and biology, vol 733. Springer, Dordrecht, pp 125–133. doi:10.1007/978-94-007-2555-3_12

Tejero R, Anitua E, Orive G (2014) Toward the biomimetic implant surface: biopolymers on titanium-based implants for bone regeneration. Prog Polym Sci 39(7):1406–1447. http://dx.doi.org/10.1016/j.progpolymsci.2014.01.001

Thakkar V, Ghobrial GM, Maulucci CM, Singhal S, Prasad SK, Harrop JS, Vaccaro AR, Behrend C, Sharan AD, Jallo J (2014) Nasal MRSA colonization: impact on surgical site infection following spine surgery. Clin Neurol Neurosurg 125:94–97. http://dx.doi.org/10.1016/j.clineuro.2014.07.018

Thomson KS, Korte FS, Giachelli CM, Ratner BD, Regnier M, Scatena M (2013) Prevascularized microtemplated fibrin scaffolds for cardiac tissue engineering applications. Tissue Eng A 19(7-8):967–977. doi:10.1089/ten.tea.2012.0286

Tomasinsig L, Morgera F, Antcheva N, Pacor S, Skerlavaj B, Zanetti M, Tossi A (2009) Structure dependence of biological activities for primate cathelicidins. J Pept Sci 15(9):576–582. doi:10.1002/psc.1143

Totsika M, Kostakioti M, Hannan TJ, Upton M, Beatson SA, Janetka JW, Hultgren SJ, Schembri MA (2013) A FimH inhibitor prevents acute bladder infection and treats chronic cystitis caused by multidrug-resistant uropathogenic Escherichia coli ST131. J Infect Dis 208(6):921–928. doi:10.1093/infdis/jit245

Triantafillopoulos I, Papaioannou N (2014) The effect of pharmacological agents on the bone-implant interface. In: Karachalios T (ed) Bone-implant interface in orthopedic surgery. Springer, London, pp 221–237. doi:10.1007/978-1-4471-5409-9_16

Tu Q, Wang J-C, Liu R, He J, Zhang Y, Shen S, Xu J, Liu J, Yuan M-S, Wang J (2013) Antifouling properties of poly(dimethylsiloxane) surfaces modified with quaternized poly(dimethylaminoethyl methacrylate). Colloids Surf B: Biointerfaces 102:361–370. http://dx.doi.org/10.1016/j.colsurfb.2012.08.033

Tuomanen E, Cozens R, Tosch W, Zak O, Tomasz A (1986) The rate of killing of Escherichia coli byβ-lactam antibiotics is strictly proportional to the rate of bacterial growth. J Gen Microbiol 132(5):1297–1304. doi:10.1099/00221287-132-5-1297

Veiga AS, Sinthuvanich C, Gaspar D, Franquelim HG, Castanho MARB, Schneider JP (2012) Arginine-rich self-assembling peptides as potent antibacterial gels. Biomaterials 33(35):8907–8916. http://dx.doi.org/10.1016/j.biomaterials.2012.08.046

Veronesi B, Chorley B, Ward W, Simmons SO, Tennant A, Vallanat B (2014) The physicochemistry of capped nanosilver predicts its biological activity in rat brain endothelial cells (RBEC4). ACS Sustain Chem Eng 2(7):1566–1573. doi:10.1021/sc5000896

Vetten MA, Yah CS, Singh T, Gulumian M (2014) Challenges facing sterilization and depyrogenation of nanoparticles: effects on structural stability and biomedical applications. Nanomed Nanotechnol Biol Med 10(7):1391–1399. http://dx.doi.org/10.1016/j.nano.2014.03.017

Wang Z, Li N, Zhao J, White JC, Qu P, Xing B (2012) CuO nanoparticle interaction with human epithelial cells: cellular uptake, location, export, and genotoxicity. Chem Res Toxicol 25(7):1512–1521. doi:10.1021/tx3002093

Wang Z, Chen Z, Zuo Q, Song F, Wu D, Cheng W, Fan W (2013) Reproductive toxicity in adult male rats following intra-articular injection of cobalt–chromium nanoparticles. J Orthop Sci 18(6):1020–1026. doi:10.1007/s00776-013-0458-2

Wang L, Su B, Cheng C, Ma L, Li S, Nie S, Zhao C-S (2014) Layer by Layer assembly of sulfonic poly(ether sulfone) as heparin-mimicking coatings: scalable fabrication of super-hemocompatible and antibacterial membranes. J Mater Chem B. doi:10.1039/C4TB01865F

Wang R, Neoh KG, Kang E-T (2015) Integration of antifouling and bactericidal moieties for optimizing the efficacy of antibacterial coatings. J Colloid Interface Sci 438:138–148. http://dx.doi.org/10.1016/j.jcis.2014.09.070

Weissman C, Murray WB (2013) It's not just another room…. Anesth Analg 117(2):287–289 210.1213/ANE.1210b1013e3182991c3182990e

Wong KKY, Liu X (2010) Silver nanoparticles-the real "silver bullet" in clinical medicine? MedChemComm 1(2):125–131. doi:10.1039/C0MD00069H

Wu F, Meng G, He J, Wu Y, Wu F, Gu Z (2014) Antibiotic-loaded chitosan hydrogel with superior dual functions: antibacterial efficacy and osteoblastic cell responses. ACS Appl Mater Interfaces 6(13):10005–10013. doi:10.1021/am502537k

Xue Y, Hieda Y, Kimura K, Takayama K, Fujihara J, Tsujino Y (2004) Kinetic characteristics and toxic effects of benzalkonium chloride following intravascular and oral administration in rats. J Chromatogr B 811(1):53–58. http://dx.doi.org/10.1016/j.jchromb.2004.03.075

Xue Y, Zhang S, Tang M, Zhang T, Wang Y, Hieda Y, Takeshita H (2012) Comparative study on toxic effects induced by oral or intravascular administration of commonly used disinfectants and surfactants in rats. J Appl Toxicol 32(7):480–487. doi:10.1002/jat.1662

Yang C, Ding X, Ono RJ, Lee H, Hsu LY, Tong YW, Hedrick J, Yang YY (2014) Brush-like polycarbonates containing dopamine, cations, and PEG providing a broad-spectrum, antibacterial, and antifouling surface via one-step coating. Adv Mater 26(43):7346–7351. doi:10.1002/adma.201402059

Ye Q, Zhou F (2015) Antifouling surfaces based on polymer brushes. In: Zhou F (ed) Antifouling surfaces and materials. Springer, Berlin, pp 55–81. doi:10.1007/978-3-662-45204-2_3

Ye J, He J, Wang C, Yao K, Gou Z (2014) Copper-containing mesoporous bioactive glass coatings on orbital implants for improving drug delivery capacity and antibacterial activity. Biotechnol Lett 36(5):961–968. doi:10.1007/s10529-014-1465-x

Yeung AY, Gellatly S, Hancock RW (2011) Multifunctional cationic host defence peptides and their clinical applications. Cell Mol Life Sci 68(13):2161–2176. doi:10.1007/s00018-011-0710-x

Yin LM, Edwards MA, Li J, Yip CM, Deber CM (2012) Roles of hydrophobicity and charge distribution of cationic antimicrobial peptides in peptide-membrane interactions. J Biol Chem 287(10):7738–7745. doi:10.1074/jbc.M111.303602

Yu Q, Zhang Y, Wang H, Brash J, Chen H (2011) Anti-fouling bioactive surfaces. Acta Biomater 7(4):1550–1557. http://dx.doi.org/10.1016/j.actbio.2010.12.021

Yu X, Ning Z, Hua M, Wang C, Cui F (2013) Mechanical and biomedical properties of copper-containing diamond-like carbon films on magnesium alloys. J Mater Chem B 1(37):4773–4780. doi:10.1039/C3TB20570C

Zaiou M, Nizet V, Gallo RL (2003) Antimicrobial and protease inhibitory functions of the human cathelicidin (hCAP18/LL-37) prosequence. J Investig Dermatol 120(5):810–816

Zasloff M (2002) Antimicrobial peptides of multicellular organisms. Nature 415(6870):389–395

Zhang X, Oglęcka K, Sandgren S, Belting M, Esbjörner EK, Nordén B, Gräslund A (2010) Dual functions of the human antimicrobial peptide LL-37—target membrane perturbation and host cell cargo delivery. Biochim Biophys Acta Biomembr 1798(12):2201–2208. http://dx.doi.org/10.1016/j.bbamem.2009.12.011

Zhao K, Tseng BS, Beckerman B, Jin F, Gibiansky ML, Harrison JJ, Luijten E, Parsek MR, Wong GCL (2013a) Psl trails guide exploration and microcolony formation in Pseudomonas aeruginosa biofilms. Nature 497(7449):388–391. doi:10.1038/nature12155. http://www.nature.com/nature/journal/v497/n7449/abs/nature12155.html#supplementary-information

Zhao Y, Chen Z, Chen Y, Xu J, Li J, Jiang X (2013b) Synergy of non-antibiotic drugs and pyrimidinethiol on gold nanoparticles against superbugs. J Am Chem Soc 135(35):12940–12943. doi:10.1021/ja4058635

Zhao Z, Ni H, Han Z, Jiang T, Xu Y, Lu X, Ye P (2013c) Effect of surface compositional heterogeneities and microphase segregation of fluorinated amphiphilic copolymers on antifouling performance. ACS Appl Mater Interfaces 5(16):7808–7818. doi:10.1021/am401568b

Zheng J, Song W, Huang H, Chen H (2010) Protein adsorption and cell adhesion on polyurethane/Pluronic® surface with lotus leaf-like topography. Colloids Surf B: Biointerfaces 77(2):234–239. http://dx.doi.org/10.1016/j.colsurfb.2010.01.032

Zhou C, Li P, Qi X, Sharif ARM, Poon YF, Cao Y, Chang MW, Leong SSJ, Chan-Park MB (2011) A photopolymerized antimicrobial hydrogel coating derived from epsilon-poly-l-lysine. Biomaterials 32(11):2704–2712. http://dx.doi.org/10.1016/j.biomaterials.2010.12.040

Zhou Z, Calabrese DR, Taylor W, Finlay JA, Callow ME, Callow JA, Fischer D, Kramer EJ, Ober CK (2014) Amphiphilic triblock copolymers with PEGylated hydrocarbon structures as environmentally friendly marine antifouling and fouling-release coatings. Biofouling 30(5):589–604. doi:10.1080/08927014.2014.897335

Zhu L-J, Zhu L-P, Yi Z, Jiang J-H, Zhu B-K, Xu Y-Y (2013) Hemocompatible and antibacterial porous membranes with heparinized copper hydroxide nanofibers as separation layer. Colloids Surf B: Biointerfaces 110:36–44. http://dx.doi.org/10.1016/j.colsurfb.2013.04.020

Chapter 8
Development of Fimbrolides, Halogenated Furanones and their Derivatives as Antimicrobial Agents

Kitty K.K. Ho*, Samuel K. Kutty*, Daniel Chan, Renxun Chen, Mark D.P. Willcox, and Naresh Kumar

Abstract Traditional treatment for bacterial infection is based on compounds that target bacterial viability or growth. However, a major concern with this approach is the frequent development of drug-resistant mutants. The discovery of bacterial quorum sensing (QS) systems, which control fundamental processes involved in bacterial physiology and virulence, has opened new avenues for the development of antimicrobial agents for the control of bacterial infections. Fimbrolides isolated from Australian native marine alga *Delisea pulchra* are excellent examples of QS inhibitors provided by nature. Fimbrolides and their analogues exhibit excellent QS inhibitory activity without interfering with bacterial growth, and thus offer promising targets for development of new strategies to control microbial colonisation of surfaces. This chapter describes the types of natural fimbrolides, their biosynthesis, and synthesis of related halogenated furanone and dihydropyrrolone analogues, as well as their biological activities and applications as antimicrobial coatings for the prevention of bacterial infections.

Keywords Marine natural products • Halogenated compounds • Toxicity • Fimbrolides • Halogenated furanones • *Delisea pulchra* • Biosynthesis • Lactam analogues

Interest in marine natural products has been intensified by their associated array of biological activities and therefore their potential biomedical applications (Newman and Cragg 2004; Konig and Wright 1996; Carte 1996; Claeson and Bohlin 1997;

*Author contributed equally with all other contributors.

K.K.K. Ho • S.K. Kutty • D. Chan • R. Chen • N. Kumar (✉)
School of Chemistry, University of New South Wales, Sydney, NSW 2052, Australia
e-mail: n.kumar@unsw.edu.au

M.D.P. Willcox (✉)
School of Optometry and Vision Science, University of New South Wales, Sydney, NSW 2052, Australia
e-mail: m.willcox@unsw.edu.au

E.P. Ivanova, R.J. Crawford (eds.), *Antibacterial Surfaces*,
DOI 10.1007/978-3-319-18594-1_8

Plavsic 2004; Faulkner 2002; Blunt et al. 2004; Amsler et al. 2001). In marine organisms, these products play a vital role in chemical defense mechanisms (Kubanek et al. 2003). Defensive functions attributed to marine secondary metabolites include predator deterrence (Vervoort et al. 1997), antifouling (Bhosale et al. 2002), inhibition of overgrowth (Pawlik 1993) and UV radiation protection (Edreva et al. 2008).

Marine natural products are remarkable in their ability to produce halogenated compounds when compared to their terrestrial counterparts (Gribble 2004). Fimbrolides or halogenated furanones are one such class of marine natural product, first isolated from the red alga *Delisea pulchra*. Fimbrolides are a class of organic compounds comprising a five-membered lactone ring and halogen substituents. They reside in vesicles on the surface of the algae and beckerilides against the fouling of algal surfaces by marine organisms. The fimbrolides have become a subject of interest as novel antimicrobial agents, especially with a the increasing microbial resistance to traditional antibiotics. This review will cover the role of natural fimbrolides and their biosynthesis, as well as the synthesis of analogues, their activities and applications of fimbrolides and their analogues as antimicrobial agents.

8.1 Fimbrolides in Nature

Fimbrolides are a group of rarely occurring metabolites extracted from sea organisms. It is a term given to describe the naturally occurring halogenated furanones, which were first reported in 1977 by two research groups (Kazlauskas et al. 1977; Pettus et al. 1977). Fimbrolides have a 4-halo-3-butyl-5-halomethylene-2-(5*H*)-furanone skeleton and were first isolated from the marine red alga *Delisea pulchra* (Greville) Montage (Class- Rhodophyta, Order-Bonnemaisonales, Family-Bonnemaisoniaceae) (cf. *Delisea fimbriata*). This alga is distributed from Southern Queensland to Tasmania in Australia and through the Antarctic Islands and Antarctic Peninsula (Bonin and Hawkesf 1988). *D. pulchra* produces more than 20 of these fimbrolides with structural variation based primarily on variable substitution at two positions, in particular a hydrogen, an acetate or hydroxyl groups at C-1′ and an exocyclic bromomethylene or dibromomethylene at C-5 (Kazlauskas et al. 1977; de Nys 1993). A large number of fimbrolides and related beckerilides, which lack the exocyclic halomethylene double bond but contain a 5-halomethyl group instead, have been described (Ohta 1977; McCombs et al. 1988; de Nys et al. 1992) (Fig. 8.1).

In addition, dimeric fimbrolides have been reported to have been obtained from algal species such as *Delisea elegans* and *D. pulchra* (Fig. 8.2). These dimeric forms usually contain brominated butane linkers to bridge the fimbrolides (McCombs et al. 1988; Ankisetty et al. 2004).

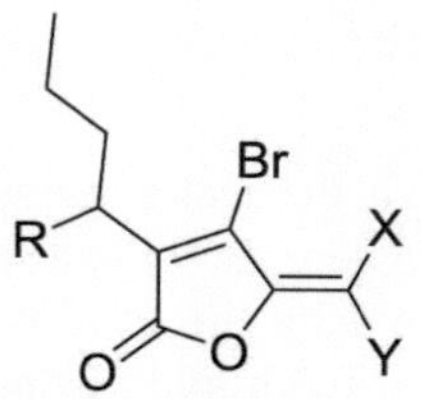

Fimbrolides

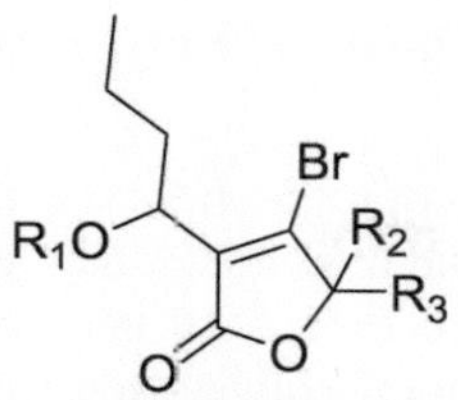

Beckerilides

1. R = H, X = Br, Y = H
2. R = H, X = H, Y = Br
3. R = H, X = Br, Y = Br
4. R = OH, X = Br, Y = H
5. R = OH, X = H, Y = Br
6. R = OH, X = I, Y = H
7. R = OH, X = H, Y = I
8. R = OH, X = Cl, Y = H
9. R = OH, X = H, Y = Cl
10. R = OH, X = Br, Y = Br
11. R = OAc, X = Br, Y = H
12. R = OAc, X = H, Y = Br
13. R = OAc, X = H, Y = I
14. R = OAc, X = I, Y = H
15. R = OAc, X = Cl, Y = H
16. R = OAc, X = H, Y = Cl
17. R = OAc, X = Br, Y = Br

18. R_1 = Ac, R_2 = OCH_3, R_3 = CH_2I
19. R_1 = H, R_2 = OCH_3, R_3 = CH_2I
20. R_1 = H, R_2 = CH_2I, R_3 = OCH_3
21. R_1 = Ac, R_2 = CH_3, R_3 = OCH_3
22. R_1 = Ac, R_2 = OCH_3, R_3 = CH_3
23. R_1 = Ac, R_2 = $CHBr_2$, R_3 = OCH_3
24. R_1 = Ac, R_2 = OCH_3, R_3 = $CHBr_2$
25. R_1 = H, R_2 = OH, R_3 = CH_3

Fig. 8.1 Chemical structures of fimbrolides and beckerilides

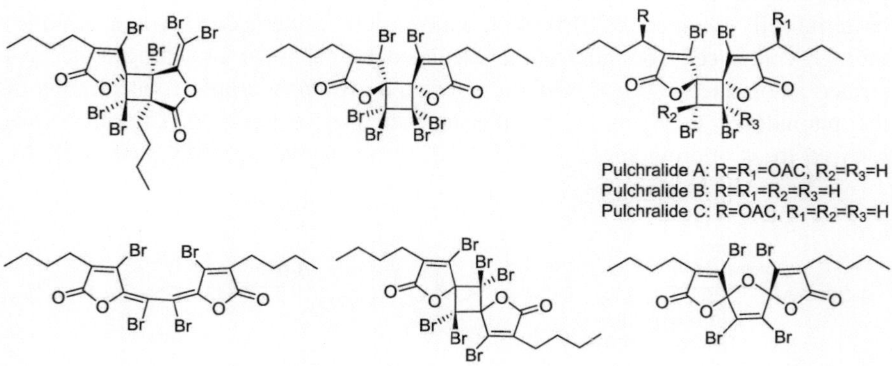

Fig. 8.2 Chemical structures of dimeric fimbrolides

8.2 Biosynthesis of Fimbrolides in Nature

The biosynthesis of fimbrolides is expected to occur via the acetate pathway, where acetyl-CoA in the presence of acyl carrier protein undergoes the claisen reaction (Dewick 2002) to form C-10 fatty acids. These fatty acids are then oxidized and halogenated to form polyhalogenated 2,4-nonadiones (with hydroxylation at C-6 for some of the lactones). Further reaction via the Favorsky rearrangement yields the polyhalogenated C-4 carboxylate substituted *n*-octane, which through enolization of the remaining carbonyl toward C-l, are cyclized to give the lactones (fimbrolides) (McConnell and Fenical 1980; Garson 1989) (Fig. 8.3).

8.3 Biological Role of Naturally Occurring Fimbrolides

Delisea species have lower bacterial accumulation on their surfaces compared to other algal species in their natural marine environment. This difference in bacterial density between the *Delisea* and other algae was later attributed to the fimbrolide compounds, which inhibit the fouling (colonising) of bacteria on their thallus surfaces (Gram et al. 1996). Fimbrolides produced by the algae are delivered to the surface at a concentration where they can regulate bacterial colonisation and the settlement of epibiota (de Nys et al. 2006a). Two parent antifouling compounds have been successfully identified from *D. pulchra*. These are 4-bromo-3-butyl-5-(dibromoethylene)-2(5*H*)-furanone and 4-bromo-5-(bromomethylene)-3-butyl-2(5*H*)-furanone (Rasmussen et al. 2000).

Under appropriate conditions, bacteria can adhere to surfaces and form biofilms. Biofilms consist of an extracellular hydrated matrix of polysaccharides and proteins produced by the bacteria themselves or acquired from the environment, and encased bacterial cells (Otter et al. 2015). Within the biofilm, bacteria control their activities through cell-to-cell communication, mediated by chemical signalling molecules termed autoinducers. When bacteria perceive a certain threshold concentration of the autoinducer, they respond by altering their gene expression. This process is referred to as quorum sensing (QS), and is responsible for many pathologically

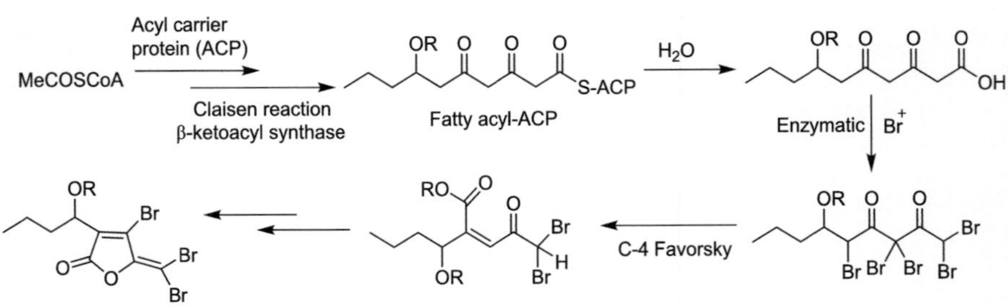

Fig. 8.3 Biosynthesis of fimbrolides

Fig. 8.4 Structures of AI-1 autoinducers (BHL and OHHL) and an example of fimbrolide

relevant processes such as virulence factor production, biofilm formation and drug resistance (Nasser and Reverchon 2007; Ni et al. 2008; Holm and Vikström 2014).

QS systems are classified according to the structure of the responsible autoinducers. Three major classes of QS systems have been identified: the *N*-acyl-L-homoserine lactone (AHL)-mediated system (AI-1), which is present in most Gram-negative bacteria; peptide-mediated QS system, employed by Gram-positive bacteria; and the autoinducer-2 (AI-2) system that utilises a family of interconvertible furanone compounds such as 4,5-dihydroxy-2,3-pentanedione as signalling molecules, and is present in both Gram-negative and Gram-positive bacteria (Williams et al. 2007).

It has been shown that natural furanones can control multicellular behaviour induced by AHL and AI-2 in Gram-negative microorganisms (Manefield et al. 1999, 2000, 2002; Ren et al. 2001, 2004a). Fimbrolides and related furanones have structural similarities to the QS autoinducers AHL, such as *N*-butanoyl-L-homoserine lactone (BHL) and *N*-3-(oxo-hexanoyl)-L-homoserine lactone (OHHL) (Fig. 8.4; AI-1 autoinducers). In the presence of fimbrolides, the cellular communication of the bacteria is inhibited. This inhibition has been demonstrated to occur as a result of competitive binding of the fimbrolides to the AHL-binding site in the QS receptor protein such as LuxR, displacing the native AHL autoinducer (Manefield et al. 1999, 2002). This leads to the disruption in the virulence related activities of microbes such as biofilm formationand the production of virulence factors (Davies et al. 1998).

For example, Defoirdt et al. studied the mechanism of QS inhibition by natural furanone (*5Z*)-4-bromo-5-(bromomethylene)-3-butyl-2(*5H*)-furanone in *Vibrio harveyi* (Defoirdt et al. 2007). Their studies indicated that the furanone disrupted the QS in *V. harveyi* by obstructing the ability of the QS master regulator protein LuxR to bind to the target promoter sequences.

The mechanism of action of halogenated furanones on a plant pathogen, *Erwinia carotovora*, has also been studied (Manefield et al. 2001). This pathogen regulates theinhibited the growth expression of virulence factors and antibiotic production via an OHHL-dependent QS mechanism. Using DNA microarrays, 79 % of the *E. coli* genes were repressed by (5*Z*)-4-bromo-5-(bromomethylene)-3-butyl-2(5*H*)-furanone and were also induced by AI-2, suggesting that the naturally occurring furanones can inhibit this system (Ren et al. 2004a).

Natural furanones also inhibit growth, swarming motility and biofilm formation in the Gram-positive bacterium *Bacillus subtilis* (Ren et al. 2002, 2004b). *Bacillus anthracis* is a Gram-positive bacterium that is the etiological agent of anthrax. Jones et al. studied the effects on *Bacillus anthracis* growth and virulence gene (*pagA, lef,* and *cya*) expression of the naturally occurring QS inhibitor *(5Z)*-4-bromo-5-(bromomethylene)-3-butyl-2*(5H)*-furanone (Jones et al. 2005). This furanone inhibited the growth of *B. anthracis* in a dose-dependent manner. Exposure to the furanone did not lead to the selection of resistant cells. This naturally occurring furanone also significantly reduced the expression of the genes over and above any effects on bacterial growth rate/viability.

8.4 Stability and Toxicity of the Naturally Occurring Fimbrolides

Despite the promising inhibitory effects of furanones on microbial biofilm formation, some of the naturally occurring furanones are unstable (Rasmussen and Givskov 2006), and there have been reports on the effects of furanones on eukaryotic cells (viability and toxicity) (Hentzer and Givskov 2003; Kuehl et al. 2009; Kjelleberg et al. 1999; Lowery et al. 2009). Hentzer et al. found that QS inhibitive furanones are very reactive and therefore speculated that they will be too toxic for therapeutic use (Hentzer and Givskov 2003). Kuehl et al. demonstrated that a natural furanone was toxic against eukaryotic L929 fibroblasts at 28 μM (Kuehl et al. 2009) and a similar value was reported by Lowery et al. (2009), while Kjelleberg et al. reported the same compound to be toxic to fibroblast cells at 483 μM (Kjelleberg et al. 1999). Another natural furanone, of similar structure but with an acetoxy group substituted at the hydrocarbon chain, was found to be much more toxic compared to the one without the substitution (Kjelleberg et al. 1999). The cytotoxicity of the furanones is unpredictable. This unpredictable behavior is structure-dependent, with slight structural variation having the ability to immensely affect their activity and toxicity.

8.5 Synthesis of Fimbrolides and Their Analogues

The first synthesis of 4-bromo-3-butyl-5-bromomethylene-2(5*H*)-furanone involved the acid-catalyzed cyclization of brominated 2-butyllevulinic acid, a remarkable reaction because of its success under the seemingly harsh conditions of concentrated sulfuric acid and elevated temperatures (Beechan and Sims 1979) (Scheme 8.1). Later re-investigation of this reaction revealed that significant amounts of the isomeric 3-butyl-5-dibromomethylene-2(5*H*)-furanone and tribromo compound 4-bromo-3-butyl-5-dibromomethylene-2(5*H*)-furanone were also formed in this reaction (Manny et al. 1997). An alternative synthesis employed the reaction of β-bromo-β-lithioacrylate with acetic anhydride to form

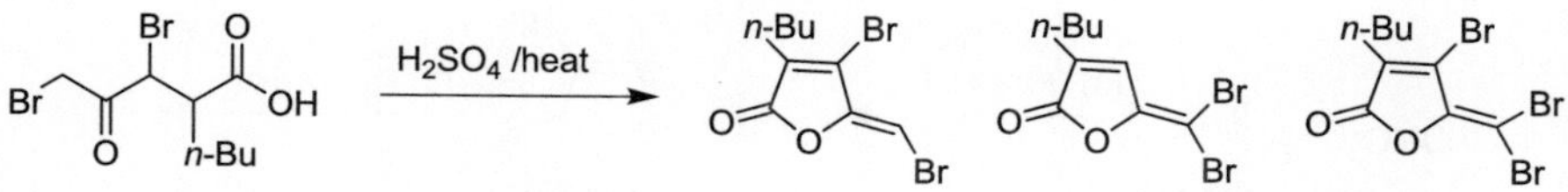

Scheme 8.1 Synthesis of 4-bromo-3-butyl-5-bromomethylene-2(5H)-furanone by acid-catalyzed cyclization of brominated 2-butyllevulinic acid

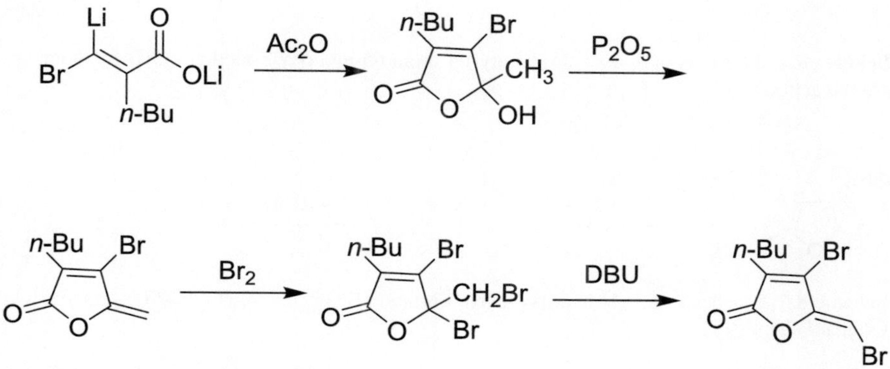

Scheme 8.2 Synthesis of 4-bromo-3-butyl-5-bromomethylene-2(5H)-furanone from β-bromo-β-lithioacrylate and acetic anhydride

Scheme 8.3 Synthesis of 4-bromo-3-butyl-5-bromomethylene-2-(5H)-furanone from N-arylmaleimide

5-hydroxy-5-methyl-4-bromo-3-butyl-2(5*H*)-furanone, which could be converted into the desired 4-bromo-3-butyl-5-bromomethylene-2(5*H*)-furanone via sequential dehydration, bromination and dehydrobromination steps (Caine and Ukachukwu 1985) (Scheme 8.2).

A more recent methodology made use of *N*-arylmaleimide as a starting material to generate the key butylmaleic anhydride intermediate, which was subjected to Grignard conditions followed by the standard transformation to yield the target fimbrolide in moderate overall yield (Haval and Argade 2007) (Scheme 8.3).

The synthesis of fimbrolides bearing oxygen functionality in the alkyl side chain is more difficult. Early efforts were directed at the incorporation of an oxygenated butyl side chain via aldol or Grignard methodologies into an appropriately substituted furan to furnish 3-(1′-hydroxybutyl)-4-bromo-5-hydroxy-5-methyl-2(5*H*)-furanone (Jefford et al. 1989) (Scheme 8.4) and 3-(1′-acetoxybutyl)-5-hydroxy-5-methyl-2(5*H*)-furanone (Kotsuki et al. 1983) (Scheme 8.5), respectively.

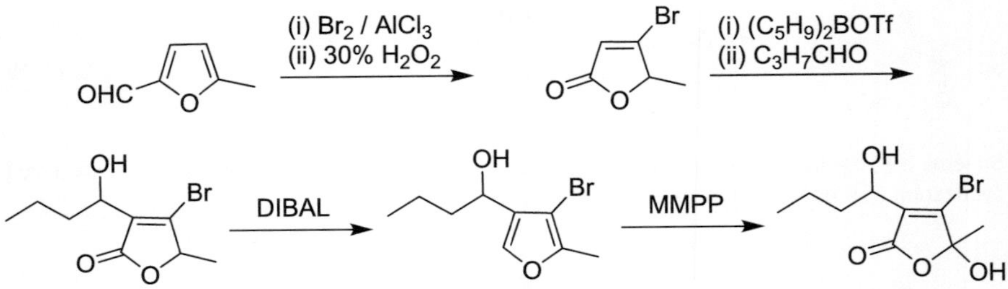

Scheme 8.4 Synthesis of 3-(1′-hydroxybutyl)-4-bromo-5-hydroxy-5-methyl-2(5H)-furanone via aldol reaction

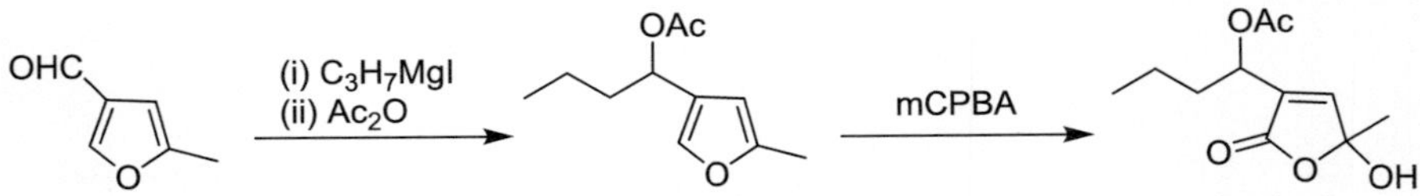

Scheme 8.5 Synthesis of 3-(1′-hydroxybutyl)-4-bromo-5-hydroxy-5-methyl-2(5H)-furanone via Grignard chemistry

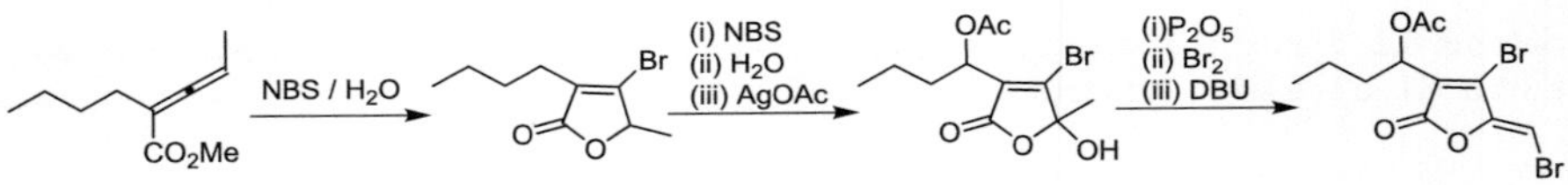

Scheme 8.6 Synthesis of 1′-acetoxyfimbrolide

These compounds may be similarly transformed into the corresponding 5-bromomethylene compounds using the methodology described above.

More recent methodologies targeted the side chain functionalization of the parent 3-butyl-2(5*H*)-furanone molecule. An elegant synthesis to the oxygenated fimbrolides involved the cyclisation of the γ-monosubstituted allenic ester with *N*-bromosuccinimide (NBS) to generate 4-bromo-3-butyl-5-methyl-2(5*H*)-furanone. This intermediate was treated again with NBS and silver acetate to produce 3-(1′-acetoxybutyl)-5-hydroxy-5-methyl-2(5*H*)-furanone, followed by the standard transformation to yield the desired 1′-acetoxyfimbrolide (Font et al. 1990; Demarch et al. 1995) (Scheme 8.6). Alternatively, the functionalization may be performed after the 5-bromomethylene moiety had been introduced (Read and Kumar 1999) (Scheme 8.7). These compounds can be further modified at C1′ to generate a variety of new analogues.

Desbutyl analogues of fimbrolides were efficiently synthesized by the condensation of an arylacetone with glyoxylic acid to form the 3-aryl-4-oxo-2-alkenoic acid intermediate, which could be transformed in three steps to the 4-aryl-3-bromo-5-methylene-2(5*H*)-furanone (Kumar and Read 2004) (Scheme 8.8). Interestingly, condensation with 4,4-dimethyl-2-pentanone gave the C1 regioisomer instead, which could be brominated and cyclized under the appropriate conditions to yield either the 3-bromo- or 4-bromo-5-alkylidene-2(5*H*)-furanone (Zhang et al. 2009,

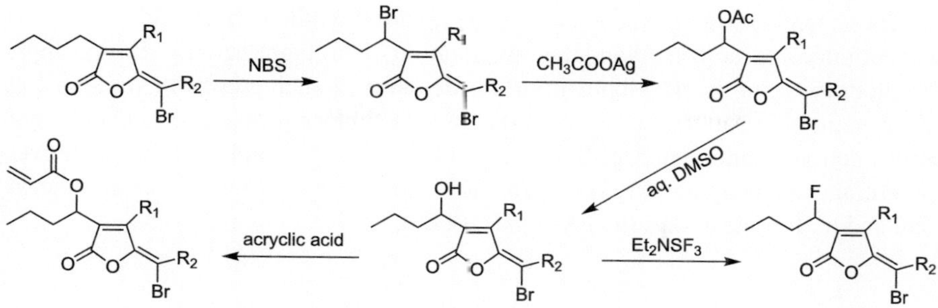

Scheme 8.7 Side chain functionalization of 3-butyl-2(5H)-furanone molecule

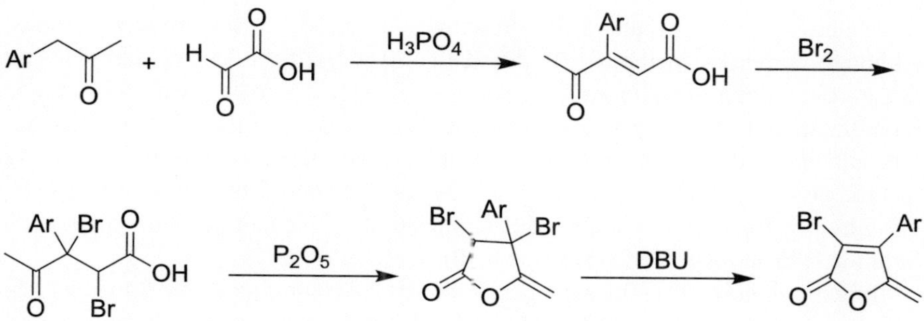

Scheme 8.8 Synthesis of 4-aryl-3-bromo-5-methylene-2(5H)-furanone

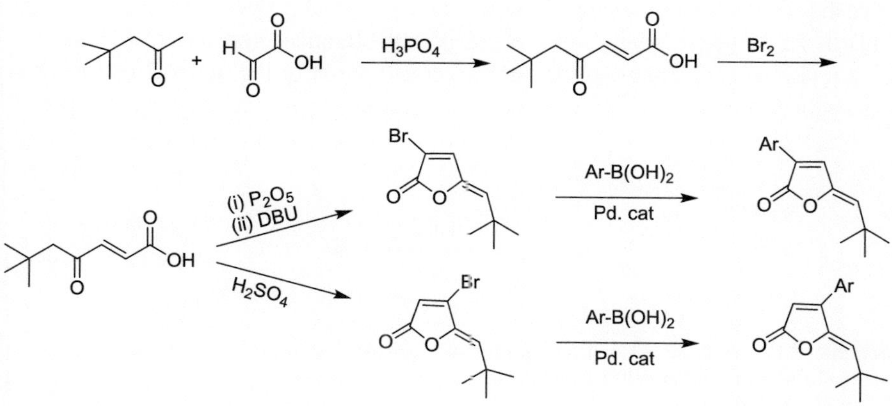

Scheme 8.9 Synthesis of 3-bromo- or 4-bromo-5-alkylidene-2(5H)-furanone, and further reactions under Suzuki-Miyaura condition to generate 3-aryl and 4-aryl-fimbrolide analogues

2011). These compounds were subjected to the Suzuki-Miyaura reaction to generate 3-aryl and 4-aryl-fimbrolide analogues (Scheme 8.9). Similarly, new 4-aryl and heteroaryl substituted furanones could be synthesized using 5-bromomethylene- and 4-bromo-5-bromomethylene-2(5*H*)-furanones as the starting compounds (Zhang et al. 2011).

These syntheses aim to generate a broad range of synthetic fimbrolide analogues for antimicrobial screening and the investigation of structure-activity relationships. Methods exist for the development of a wide range of analogues with variations in the bromination pattern, substitution on the 5-methylene group, alkyl chain length and functionalization, and oxidation level of the furanone molecule. To date, over 300 furanone analogues have been generated and screened in a variety of antimicrobial and biofouling assays (de Nys et al. 2006b).

8.6 Synthesis of Lactam Analogues

A related structure to the 2(5*H*)-furanones is the 1, 5-dihydropyrrol-2-ones. Although numerous methods exist for the generation of the lactam scaffold, few target the direct analogues of the fimbrolides, namely, the 4-bromo-5-bromomethylene-1,5-dihydropyrrol-2-ones. The synthesis of these brominated lactam analogues therefore represents a relatively new area of research. An early methodology reported the synthesis of 1-benzyl-4-bromo-5-hydroxy-5-(*n*-heptyl)-1,5-dihydropyrrol-2-one via the halolactamization of allenamides with copper (II) bromide (Ma and Xie 2005) (Scheme 8.10).

Building on this work, the analogous 1-benzyl-4-bromo-5-hydroxy-5-methyl-1,5-dihydropyrrol-2-one was synthesized and subjected to the standard dehydration-bromination conditions to yield the 5-bromo-5-bromomethylpyrrolone (Scheme 8.11). Attempted dehydrobromination with diazabicycloundecene (DBU), however, failed to give the desired 5-bromomethylene compound, presumably due to the increased steric bulk of the *N*-benzyl moiety which is absent in the furanone system (Goh 2008).

An alternative strategy to the de novo construction of the pyrrol-2-one nucleus via cyclization reactions is the lactone-lactam conversion. Recently, it was reported

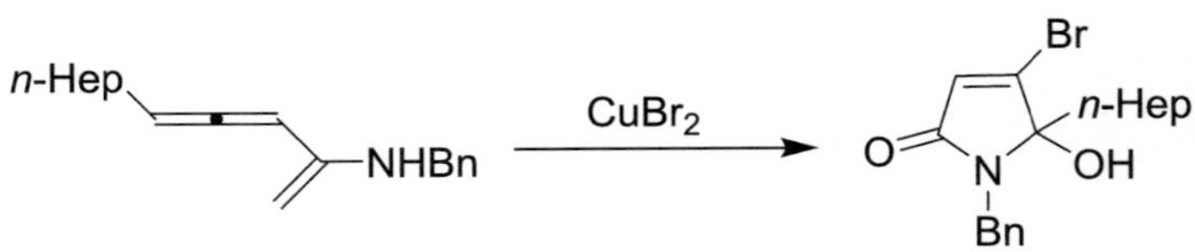

Scheme 8.10 Synthesis of 1-benzyl-4-bromo-5-hydroxy-5-(n-heptyl)-1,5-dihydropyrrol-2-one via the halolactamization of allenamides with copper (II) bromide

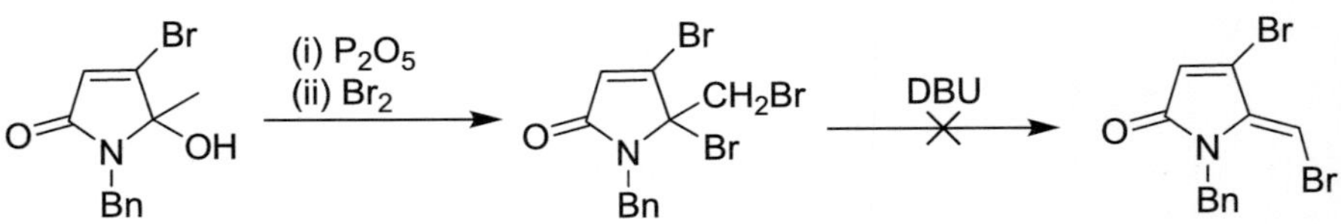

Scheme 8.11 Synthesis of 1-benzyl-4-bromo-5-bromo-5-bromomethylpyrrolone and attempted dehydrobromination with DBU

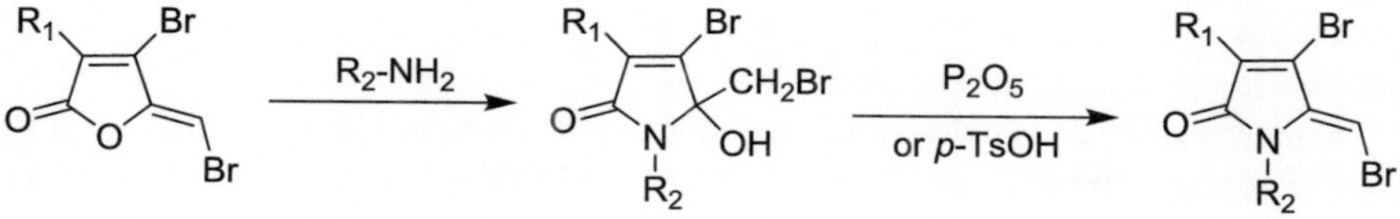

Scheme 8.12 A direct lactone to lactam conversion for the synthesis of dihydropyrrol-2-ones

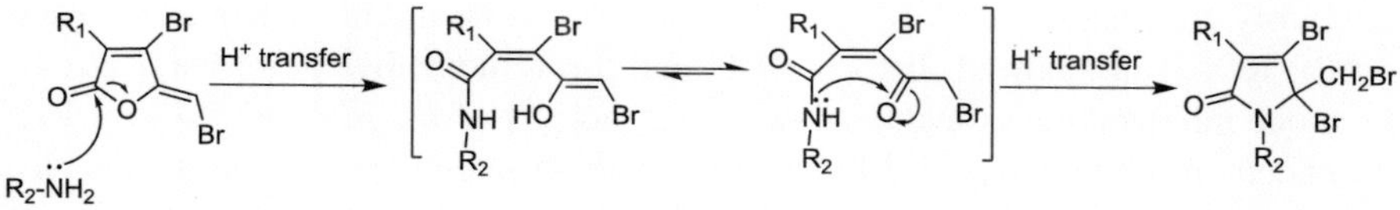

Scheme 8.13 The postulated mechanism of action of the lactone-lactam conversion

that 4-bromo-5-bromomethylene-1,5-dihydropyrrol-2-one could be efficient prepared from the lactamization of the corresponding fimbrolides, followed by dehydration (Goh et al. 2007) (Scheme 8.12).

This method circumvents the sterically demanding tribromo intermediate and the desired lactam analogues are formed in high yields. The mechanism of the lactone-lactam conversion is postulated to involve the ring-opening nucleophilic attack of the amine to the carbonyl carbon to form an amide intermediate. Subsequent keto-enol tautomerism to the keto form followed by intramolecular cyclization yields the γ-lactam ring (Goh et al. 2007) (Scheme 8.13).

The widespread occurrence of the 1,5-dihydropyrrolone system in natural products and biologically active compounds provides impetus to the continuing investigation and development of synthetic methods for this class of compound.

8.7 Bioactivity of Synthetic Analogous

In attempts to improve the safety and stability of the naturally occurring fimbrolides, several groups have embarked on the synthesis of structural analogues. Lonn-Stensrud et al. used four furanones resembling natural brominated furanones of *D. pulchra* and seven synthetic halogenated furanones (Lonn-Stensrud et al. 2009). These authors compared the ability of these compounds to prevent *Staphylococcus epidermidis* infections, and their safety. Two of the synthetic analogues, (*Z*)-5-(bromomethylene)furan-2(5*H*)-one and (*Z*)-3-bromo-5-(bromomethylene)furan-2(5*H*)-one, were able to inhibit the growth of *S. epidermidis* by interference with the AI-2 cell signalling system (Defoirdt et al. 2006; Lonn-Stensrud et al. 2007). In addition, these compounds had no irritative or genotoxic effects when tested in chicken eggs or when fed to mice. Indeed, feeding the mice for 21 days with the furanones did not alter the expression of 30,000 genes when tested on a mouse microarray (Lonn-Stensrud et al. 2009).

Kozminykh et al. tested various analogues of 3(2*H*)furanones and found varying antimicrobial activity against *S. aureus* P-209 strain (Kozminykh et al. 1993). Gein et al. used 4-aroyl-3-hydroxy-2,5-dihydrofuran-2-ones and found growth inhibitory activity against the same *S. aureus* strain and *E. coli* strain M-17 (Gein et al. 2000). Another derivative of furanones, 3-arylidene-5-(biphenyl-4-yl)-2(3*H*)-furanones, were effective as antimicrobial agents against *S. aureus* and *E. coli*, with minimum inhibitory concentrations (MIC) ranging from 10 to 100 μg/mL (Khan and Husain 2002). Ren et al. showed that (5*Z*)-4-bromo-5-(bromomethylene)-3-butyl-2(5*H*)-furanone inhibited the growth and swarming motility of *Bacillus subtilis* in a dose dependent manner (Ren et al. 2002). Additionally this compound decreased the biofilm thickness produced by this bacterium by up to 25 %, decreased the number of water channels in biofilms (showing effects on biofilm structure), and reduced the number of live cells by 63 %. Janssens et al. synthesized a small focused library of brominated furanones and tested their activity against *S. enteric* serovar Typhimurium biofilm formation (Janssens et al. 2008). *Salmonella enterica* serovar Typhimurium is a main cause of bacterial food-borne diseases. Several brominated furanones including (*Z*)-4-bromo-5-(bromomethylene)-3-alkyl-2(*5H*)-furanone had inhibitory effects on *Salmonella* biofilm formation. Interestingly, pre-treatment with the furanone rendered *Salmonella* biofilms more susceptible to antibiotic treatment. In the same study, a microarray was performed to analyse the gene expression profiles of *Salmonella* in the presence of (*Z*)-4-bromo-5-(bromomethylene)-3-ethyl-2(*5H*)-furanone and the result demonstrated interference with the synthesis of flagella by *Salmonella*.

By using DNA microarrays, Hentzer et al. found that the synthetic 4-bromo-5-(bromomethylene)-2(5*H*)-furanone repressed 85 genes of *Pseudomonas aeruginosa* PAO1, 80 % of which are also induced by AI-1 QS (Hentzer et al. 2003). Studies with a collection of structurally similar synthetic fimbrolides suggest that the conjugated exocyclic vinyl bromide on the furanone ring is the most important structural element for the non-toxic but inhibitory activity for *Escherichia coli*biofilm formation (Han et al. 2008). Furanones bearing monosubstituted bromide groups on saturated carbons were found to have a toxic effect that attenuated the bacterial growth. Fimbrolides do not affect the size of the bacterial colony, or its growth, but rather they are able to inhibit biofilm formation and the swarming motility (Ren et al. 2001; Hentzer et al. 2002). This is said to occur through their interference with the AI-2 bacterial signalling system (Ren et al. 2004a).

Synthetic derivatives of *(5Z)*-4-bromo-5-(bromomethylene)-3-butyl-2*(5H)*-furanone substantially reduced the growth of *B. anthracis* Sterne strain as well as inhibiting the expression of *pagA, lef*, or *cya* virulence genes (Jones et al. 2005). *Streptococcus* species are found on teeth, and in the gastrointestinal and genitourinal tracts. In some Streptococci, the AI-2 system is reported to be involved in virulence expression and biofilm formation (Lonn-Stensrud et al. 2007). A synthetic halogenated furanone (*Z*)-5-bromomethylene-2(5*H*)-furanone inhibited biofilm formation by *Streptococcus anginosus*, *Streptococcus intermedius*, and *Streptococcus mutans*, as well as bioluminescence induction by *Vibrio harveyi* BB152 (Lonn-

Stensrud et al. 2007). The authors suggested that the effect was linked to interference with the AI-2 signalling pathway because adding furanone to the medium had no effect on the ability of the AI-2-defective *S. anginosus luxS* and *S. intermedius luxS* mutants to form biofilms.

Fimbrolides have also been reported to have antifungal properties (Holmstrom and Kjelleberg 2001). Duo et al. showed that brominated furanones can inhibit the growth of *Candida albicans* (Duo et al. 2010). This study examined ten structurally related furanones and found that the exocyclic vinyl bromide conjugated with the carbonyl group was the most important element for fungal inhibition. Using DNA microarrays, the authors demonstrated that the synthetic furanones, *(5Z)*-4-bromo-5-(bromomethylene)-3-butyl-2*(5H)*-furanone, upregulated 32 *C. albicans genes* with functions in the stress response, NADPH dehydrogenation, and small molecule transport. However, 21 genes involved mainly in cell-wall maintenance were repressed.

Alam et al. conducted a series of bioactivity studies with 3-arylidene-5-(4-chloro-phenyl)-2(3H)-furanones and their nitrogen analogues 1-benzylpyrrolones such as anti-inflammatory, analgesic, ulcerogenic and lipid peroxidation actions (Alam et al. 2009). The studies have shown that 1-benzyl pyrrolones containing halogen groups such as 3-(4-chloro-benzylidene)-5-(4-chloro-phenyl)-1-benzyl-2(3*H*)-pyrrolone and 3-(4-fluoro-benzylidene)-5-(4-chlorophenyl)-1-benzyl-2(3*H*)-pyrrolone showed high degrees of anti-inflammatory activity using a rat paw model of oedema. Their activity was comparable to that of Ibuprofen, a non-steroidal anti-inflammatory drug used as a standard. These compounds showed an interesting profile of analgesic activity in acetic acid induced writhing test (peripheral effect; abdominal constriction in mice after injection) and in the hot-plate test (central effect; effect of increasing temperature on behaviour of mice standing on the hot-plate). The compounds were also tested for their ulcerogenic and lipid peroxidation action and showed superior gastrointestinal safety profile along with reduction in lipid peroxidation as compared to that of the Ibuprofen standard.

8.8 Applications of Halogenated Furanones

A major barrier to the long-term use of medical devices is in the development of infection. Various methods have been investigated for the binding of furanones to the surfaces of polymers used in the manufacture of medical devices. These techniques include direct surface attachment, polymerisation and physical adsorption.

The surface attachment of furanones to biomaterials can be efficiently achieved via the nitrene insertion reaction (Al-Bataineh et al. 2006). This strategy has been used to attach furanones onto catheters (Hume et al. 2004). Silastic Tenckhoff catheters were first coated with plasma-activated heptylamine vapour to generate surface amino groups. Treatment with polyacrylic acid (MW 250,000) in the presence of *N*-(3-dimethylaminopropyl)-*N*′-ethylcarbodiimide hydrochloride (EDC) gave carboxyl-terminated chains, followed by EDC coupling of 4-azidoaniline to the carboxyl moiety. Finally, nitrene insertion of 3-(1′-bromohexyl)-5-

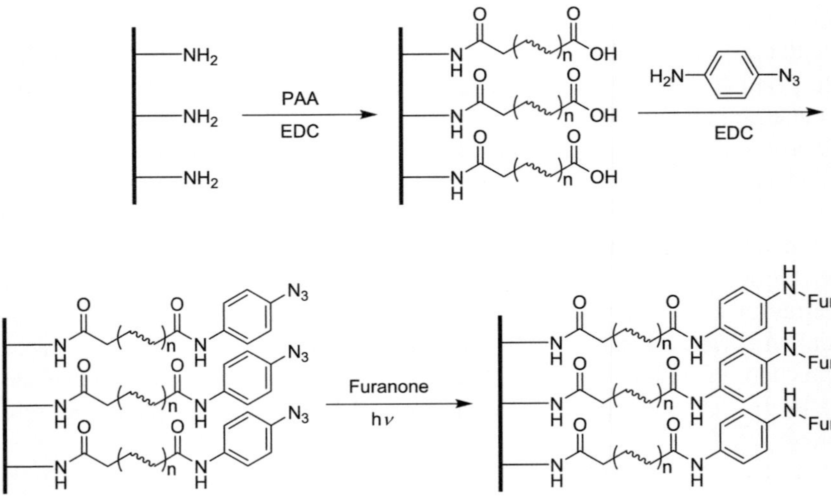

Scheme 8.14 Surface immobilization of furanones on biomaterial

dibromomethylene-2(5*H*)-furanone under UV irradiation gave the target material a surface bromine concentration of approximately 0.48 at.% (Scheme 8.14). Alternatively, a thickness controllable azide surface can be prepared by an azido silane linker, followed by nitrene insertion of furanone to give the desired furanone coating (Al-Bataineh et al. 2009).

Using an atom transfer polymerisation technique, styrene or acrylate were polymerised in the presence of a 3-(1′-bromohexyl)-5-dibromomethylene-2(5*H*)-furanone (Hume et al. 2004). This technique generates polymers with terminal furanone groups. Alternatively, bulk polymerisation uses 3-(1′-acryloyloxyalkyl)-furanone with acrylate to produce copolymers with furanones incorporated along the length of the chain (Hume et al. 2004; Baveja et al. 2004a). The covalently attached 3-(1′-bromohexyl)-5-dibromomethylene-2(5*H*)-furanone can reduce biofilm formation in *S. epidermidis* in vitro by 89 % for polystyrene coated furanones disks and 78 % by furanone-coated catheters (Hume et al. 2004).

Furanones can also be physically adsorbed directly on to surfaces. A solution of 3-(1′-bromohexyl)-5-dibromomethylene-2(5*H*)-furanone in ethanol (1 mg/mL) was applied to polymer materials and allowed to evaporate to give 40 μg/cm squared (or cm^2) (Konig and Wright 1996) of furanone. The furanone compound significantly inhibited the growth of *S. epidermidis* on the surfaces exposed to inoculum of 1×10^7 colony forming units/mL and reduced slime deposition on materials by 30–90 %, depending of the type of polymer tested (Baveja et al. 2004b). Further, the compound did not incite an acute inflammatory response either in vitro or in vivo (Baveja et al. 2004b). Their excellent biological performance, combined with their antibacterial properties, suggests that this furanone compound could be an effective anti-infective coating for implantable devices.

Zhu et al. investigated the antibacterial activity of furanone on commercially available contact lenses (Zhu et al. 2008). They found that furanone-coated lenses reduced the adhesion of bacterial strains such as *Pseudomonas aeruginosa*, *Staphylococcus aureus*, *Serratia marcescens*, and the protozoan *Acanthamoeba spp*., with a reduction of cells on surfaces between 67 and 92 %. This group was also the first to demonstrate the safety of the furanone-coated lenses in animals and humans. Guinea Pigs wore the furanone-coated contact lenses for 1 month, and showed no differences in their ocular responses; neither conjunctival redness nor corneal staining were found to be different between test and control lenses (Zhu et al. 2008). Human volunteers wore furanone-coated lenses for one day and one night. The ocular response was not different to that obtained when using uncoated lenses. The only significant difference was the subjects noticed the edge of the lenses when wearing the furanone-coated lenses more than when wearing the controls, but this may have been due to increased manipulation of lenses during the coating process rather than the furanones themselves (Zhu et al. 2008).

Another emerging application of fimbrolides is in food preservation. In their work, Shobharani and Agarwal (2010) demonstrated an increase in the shelf life of fermented milk prepared by culturing the probiotic *Leuconostoc mesenteroides* in the presence of 2(5*H*)-furanone and bromofuranone. Usually this nutritious drink is easily spoiled by the growth of a large number of microorganisms, especially *Pseudomonas spp*. Their studies showed reduced motility, reduced acyl homoserine lactone (HHSL and BHSL), low rhamnolipid content and reduced exoprotease activity of *Pseudomonas spp*. when furanone was used as a supplement.

8.9 Effects of Dihydropyrrolones (DHP)

In the process of imitating structures of antibacterial fimbrolides, a related structure, 1,5-dihydropyrrol-2-one, was also found to possess anti-QS and antibiofilm activities. The direct lactone-lactam conversion mentioned previously allows the unique bromination pattern on the fimbrolide to persist, as well as providing high yields (Goh et al. 2007). More importantly, the lactam ring system is hydrolytically more stable, making it less susceptible to lactonolysis (ring opening) in physiological conditions (Yates et al. 2002; Byers et al. 2002). A range of DHPs were successfully synthesized and has shown excellent anti-biofilm activity against *S. epidermidis*, *P. aeruginosa* and *E. coli* when present in solution (Kumar and Iskander 2007). Some of these compounds exhibited superior activity with 93 % reduction in biofilm formation of *E. coli* at concentration as low as 6 μM, with minimal cytotoxicity.

8.10 Surface Attachment of DHP as Antimicrobial Coating

The first successful strategy for the immobilization of DHP was achieved by coupling of acrylate-functionalized DHP onto an amino-functionalized surface via a Michael-type addition reaction (Fig. 8.5) (Ho et al. 2010). The DHP-coated surfaces were able to prevent the formation of biofilm by *P. aeruginosa* and *S. aureus*, and

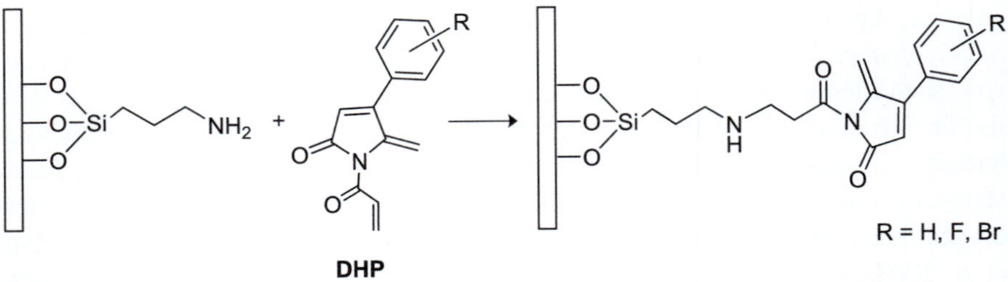

Fig. 8.5 Attachment of DHP onto an amino-functionalized surface

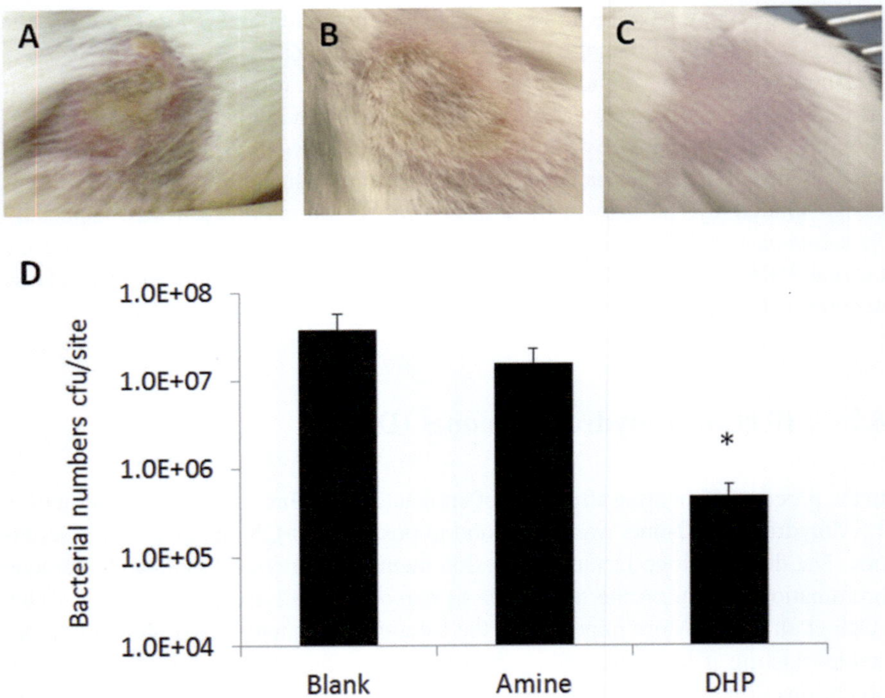

Fig. 8.6 Representative images of a mouse after 5 days of subcutaneous injection of (**a**) untreated, (**b**) process control, and (**c**) DHP polyacrylamide beads. (**d**) Average numbers of bacteria recovered from explanted biomaterials and associated tissue (Reproduced with permission from American Society for Microbiology (Ho et al. 2012))

these surfaces showed significant reductions in bacterial adhesion without increased killing levels for both strains of bacteria.

A similar attachment strategy was also applied on a polymeric bead substrate (Ho et al. 2012), and the substrate was found to be effective in reducing the number of culturable clinical isolates of *S. aureus* in a dose-dependent manner *in vitro*. Moreover, *in vivo* tests have shown that the DHP-coated beads are able to reduce the pathogenic potential and the bacterial load in a subcutaneous *S. aureus* infection model (Fig. 8.6). Importantly, the DHP beads did not cause any apparent adverse effect on the mice and appeared to be biocompatible.

The immobilization of DHP was further examined using the highly efficient click or copper-catalyzed azide-alkyne1,3-dipolar cycloaddition (CuAAC) reaction (Ho et al. 2014). The DHP was first functionalized with either an azide- or alkyne-functional group, followed by the copper-catalyzed reaction with the corresponding alkyne- or azide-functional surfaces. This strategy yield twice the surface concentration of DHP compared to the surfaces developed earlier using the Michael addition reaction. The optimized DHP coatings exhibited a significant improvement in reducing the bacterial adhesion and biofilm formation for *P. aeruginosa* and *S. aureus* cells, with observed reductions in attachment of up to 97.3 % and 96.8 % for *P. aeruginosa* and *S. aureus*, respectively (Ho et al. 2014). Importantly, these covalently bound DHPs were found to disrupt the quantum sensing (QS) of *P. aeruginosa* by interference with the AHL-mediated *las* QS system (Ho et al. 2014). This result indicated that covalently bound QS inhibitor can interact directly with the cell wall or cell membranes of bacteria, suggesting the existence of membrane QS receptors. This finding further affirms the potential of surface-bound QS inhibitors as effective antibacterial coatings.

8.11 Conclusion

The discovery of antifouling marine organisms and their secondary metabolites opens up new avenues for the development of antimicrobial agents for the control of bacterial infections. Fimbrolides, synthetic furanones and their analogues show promising activity in their ability to disrupt bacterial QS for both Gram-negative and Gram-positive bacteria at sub-bactericidal concentrations. Such a non-conventional action that acts without killing, is unlikely to induce evolutionary stress and therefore is believed to decrease the opportunity for development of resistance. Despite the promising activities of fimbrolide and their analogues, the complete mechanism of action is not yet understood, nor are their possible side effects. Indeed, if these compounds only interfere with the QS systems of microbes, then our understanding of these systems may need to be refined in order to accommodate the fact that the fimbolides and analogues are active when covalently bound to a surface. Therefore, a thorough understanding of the effects of these compounds is needed to uncover the therapeutic potential of these agents.

References

Alam MM, Husain A, Hasan SM, Suruchi, Anwer T (2009) Synthesis and pharmacological evaluation of 2(3H)-furanones and 2(3H)-pyrrolones, combining analgesic and anti-inflammatory properties with reduced gastrointestinal toxicity and lipid peroxidation. Eur J Med Chem 44:2636–2642

Al-Bataineh SA, Britcher LG, Griesser HJ (2006) XPS characterization of the surface immobilization of antibacterial furanones. Surf Sci 600:952–962

Al-Bataineh SA, Luginbuehl R, Textor M, Yan M (2009) Covalent immobilization of antibacterial furanones *via* photochemical activation of perfluorophenylazide. Langmuir 25:7432–7437

Amsler CD, McClintock JB, Baker BJ (2001) Secondary metabolites as mediators of trophic interactions among antarctic marine organisms. Am Zool 41:17–26

Ankisetty S, Nandiraju S, Win H, Park YC, Amsler CD, McClintock JB, Baker JA, Diyabalanage TK, Pasaribu A, Singh MP, Maiese WM, Walsh RD, Zaworotko MJ, Baker BJ (2004) Chemical investigation of predator-deterred macroalgae from the Antarctic peninsula. J Nat Prod 67:1295–1302

Baveja JK, Li G, Nordon RE, Hume EBH, Kumar N, Willcox MDP, Poole-Warren LA (2004a) Biological performance of a novel synthetic furanone-based antimicrobial. Biomaterials 25:5013–5021

Baveja JK, Willcox MDP, Hume EBH, Kumar N, Odell R, Poole-Warren LA (2004b) Furanones as potential anti-bacterial coatings on biomaterials. Biomaterials 25:5003–5012

Beechan CM, Sims JJ (1979) The first synthesis of fimbrolides, a novel class of halogenated lactones naturally occurring in the red seaweed *Delisea fimbriata* (Bonnemaisoniaceae). Tetrahedron Lett 20:1649–1652

Bhosale SH, Nagle VL, Jagtap TG (2002) Antifouling potential of some marine organisms from India against species of Bacillus and Pseudomonas. Mar Biotechnol (NY) 4:111–118

Blunt JW, Copp BR, Munro MH, Northcote PT, Prinsep MR (2004) Marine natural products. Nat Prod Rep 21:1–49

Bonin DR, Hawkesf MW (1988) Systematics and life histories of New Zealand Bonnemaisoniaceae (Bonnemaisoniales, Rhodophyta): II The genus *Delisea*. N Z J Bot 26:619–632

Byers JT, Lucas C, Salmond GPC, Welch M (2002) Nonenzymatic turnover of an *Erwinia carotovora* quorum-sensing signaling molecule. J Bacteriol 184:1163–1171

Caine D, Ukachukwu VC (1985) A new synthesis of 3-normal-butyl-4-bromo-5(Z)-(bromomethylidene)-2-(5H)-furanone, a naturally-occurring fimbrolide from Delisia-fimbriata (Bonnemaisoniaceae). J Org Chem 50:2195–2198

Carte BK (1996) Biomedical potential of marine natural products. Bioscience 46:271–286

Claeson P, Bohlin L (1997) Some aspects of bioassay methods in natural-product research aimed at drug lead discovery. Trends Biotechnol 15:245–248

Davies DG, Parsek MR, Pearson JP, Iglewski BH, Costerton JW, Greenberg EP (1998) The involvement of cell-to-cell signals in the development of a bacterial biofilm. Science 280:295–298

de Nys R (1993) New halogenated furanones from the marine alga *Delisea pulchra* (cf.fimbriata). Tetrahedron 49:11213–11212

de Nys R, Coll JC, Bowden BF (1992) Delisea pulchra (cf. fimbriata) revisited. The structural determination of two new metabolites from the red alga Delisea pulchra. Aust J Chem 45:1625–1632

de Nys R, Givskov M, Kumar N, Kjelleberg S, Steinberg PD (2006a) Furanones. Prog Mol Subcell Biol 42:55–86

de Nys R, Givskov M, Kumar N, Kjelleberg S, Steinberg PD (2006b) Furanones. In: Fusetani N, Clare AS (eds) Antifouling compounds. Springer, Berlin, pp 55–86

Defoirdt T, Crab R, Wood TK, Sorgeloos P, Verstraete W, Bossier P (2006) Quorum sensing-disrupting brominated furanones protect the gnotobiotic brine shrimp *Artemia franciscana* from pathogenic *Vibrio harveyi* Vibrio campbellii, and Vibrio parahaemolyticus isolates. Appl Environ Microbiol 72:6419–6423

Defoirdt T, Miyamoto CM, Wood TK, Meighen EA, Sorgeloos P, Verstraete W, Bossier P (2007) The natural furanone (5Z)-4-bromo-5-(bromomethylene)-3-butyl-2(5H)-furanone disrupts quorum sensing-regulated gene expression in Vibrio harveyi by decreasing the DNA-binding activity of the transcriptional regulator protein luxR. Environ Microbiol 9:2486–2495

Demarch P, Font J, Gracia A, Zheng QY (1995) Easy access to 5-alkyl-4-bromo-2(5h)-furanones – synthesis of a fimbrolide, an acetoxyfimbrolide, and bromobeckerelide. J Org Chem 60:1814–1822

Dewick PM (2002) Medicinal natural products: a biosynthetic approach, 2nd edn. Wiley, Chichester

Duo M, Zhang M, Luk YY, Ren D (2010) Inhibition of Candida albicans growth by brominated furanones. Appl Microbiol Biotechnol 85:1551–1563

Edreva A, Velikova V, Tsonev T, Dagnon S, Gürel A, Aktaş L, Gesheva E (2008) Stress-protective role of secondary metabolites: diversity of functions and mechanisms. Gen Appl Plant Physiol 34:67–78

Faulkner DJ (2002) Marine natural products. Nat Prod Rep 19:1–48

Font J, Gracia A, Demarch P (1990) Synthesis of 5-alkyl-4-bromo-5-hydroxy-2(5h)-furanones and 5-alkylidene-4-bromo-2(5h)-furanones. Tetrahedron Lett 31:5517–5520

Garson MJ (1989) Biosynthetic studies on marine natural products. Nat Prod Rep 6:143–170

Gein VL, Gein LF, Bezmaternykh EN, Veronina EV (2000) Synthesis and antimicrobial activity of 4-aroyl-3-hydroxy-2,5-dihydrofuran-2-ones and their derivatives. Pharm Chem J 34:254–256

Goh WK (2008) In: School of chemistry. PhD thesis, The University of New South Wales, Kensington, Sydney

Goh WK, Iskander G, Black DS, Kumar N (2007) An efficient lactamization of fimbrolides to novel 1,5-dihydropyrrol-2-ones. Tetrahedron Lett 48:2287–2290

Gram L, de Nys R, Maximilien R, Givskov M, Steinberg P, Kjelleberg S (1996) Inhibitory effects of secondary metabolites from the red alga Delisea pulchra on swarming motility of Proteus mirabilis. Appl Environ Microbiol 62:4284–4287

Gribble GW (2004) Natural organohalogens: a new fontier for medicinal agents? J Chem Educ 81:1441–1449

Han Y, Hou S, Simon KA, Ren D, Luk YY (2008) Identifying the important structural elements of brominated furanones for inhibiting biofilm formation by Escherichia coli. Bioorg Med Chem Lett 18:1006–1010

Haval, KP, Argade NP (2007) Synthesis of natural fimbrolides. Synthesis 2007:2198–2203

Hentzer M, Givskov M (2003) Pharmacological inhibition of quorum sensing for the treatment of chronic bacterial infections. J Clin Invest 112:1300–1307

Hentzer M, Riedel K, Rasmussen TB, Heydorn A, Andersen JB, Parsek MR, Rice SA, Eberl L, Molin S, Hoiby N, Kjelleberg S, Givskov M (2002) Inhibition of quorum sensing in *Pseudomonas aeruginosa* biofilm bacteria by a halogenated furanone compound. Microbiology 148:87–102

Hentzer M, Wu H, Andersen JB, Riedel K, Rasmussen TB, Bagge N, Kumar N, Schembri MA, Song Z, Kristoffersen P, Manefield M, Costerton JW, Molin S, Eberl L, Steinberg P, Kjelleberg S, Hoiby N, Givskov M (2003) Attenuation of Pseudomonas aeruginosa virulence by quorum sensing inhibitors. EMBO J 22:3803–3815

Ho KKK, Cole N, Chen R, Willcox MDP, Rice SA, Kumar N (2010) Characterisation and in vitro activities of surface attached dihydropyrrol-2-ones against Gram-negative and Gram-positive bacteria. Biofouling 26:913–921

Ho KKK, Cole N, Chen R, Willcox MDP, Rice SA, Kumar N (2012) Immobilization of antibacterial dihydropyrrol-2-ones on functional polymer supports to prevent bacterial infections in vivo. Antimicrob Agents Chemother 56:1138–1141

Ho KKK, Chen R, Willcox MDP, Rice SA, Cole N, Iskander G, Kumar N (2014) Quorum sensing inhibitory activities of surface immobilized antibacterial dihydropyrrolones via click chemistry. Biomaterials 35:2336–2345

Holm A, Vikström E (2014) Quorum sensing communication between bacteria and human cells: signals, targets, and functions. Front Plant Sci 5:309

Holmstrom C, Kjelleberg S (2001) Inhibition of fungi. PCT, patent number WO2001068091

Hume EBH, Baveja J, Muir B, Schubert TL, Kumar N, Kjelleberg S, Griesser HJ, Thissen H, Read R, Poole-Warren LA, Schindhelm K, Willcox MDP (2004) The control of Staphylococcus epidermidis biofilm formation and *in vivo* infection rates by covalently bound furanones. Biomaterials 25:5023–5030

Janssens JC, Steenackers H, Robijns S, Gellens E, Levin J, Zhao H, Hermans K, De Coster D, Verhoeven TL, Marchal K, Vanderleyden J, De Vos DE, De Keersmaecker SC (2008)

Brominated furanones inhibit biofilm formation by Salmonella enterica serovar Typhimurium. Appl Environ Microbiol 74:6639–6648

Jefford CW, Jaggi D, Boukouvalas J (1989) Total synthesis of bromobeckerelide. Tetrahedron Lett 30:1237–1240

Jones MB, Jani R, Ren D, Wood TK, Blaser MJ (2005) Inhibition of Bacillus anthracis growth and virulence-gene expression by inhibitors of quorum-sensing. J Infect Dis 191:1881–1888

Kazlauskas R, Murphy PT, Quinn RJ, Wells RJ (1977) A new class of halogenated lactones from Delisea fimbriata (Bonnemaisoniaceae). Tetrahedron Lett 18:37–40

Khan MS, Husain A (2002) Syntheses and reactions of some new 2-arylidene-4-(biphenyl-4-yl)-but-3-en-4-olides with a study of their biological activity. Pharmazie 57:448–452

Kjelleberg S, Steinberg P, Holmstrom C, Back A (1999) Inhibition of gram positive bacteria. PCT, patent number WO1999053915

Konig GM, Wright AD (1996) Marine natural products research: current directions and future potential. Planta Med 62:193–211

Kotsuki H, Monden M, Ochi M (1983) Efficient synthesis of acetoxyfimbrolides and beckerelide analogs. Chem Lett 12:1007–1008

Kozminykh VO, Igidov NM, Kozminykh EN, Aliev ZG (1993) Reactions of 5-aryl-furan-2,3-diones with acylmethylenetriphenylphosphoranes: synthesis and biological activity of 3(2H)-furanone derivatives. Pharmazie 48:99–106

Kubanek J, Jensen PR, Keifer PA, Sullards MC, Collins DO, Fenical W (2003) Seaweed resistance to microbial attack: a targeted chemical defense against marine fungi. Proc Natl Acad Sci U S A 100:6916–6921

Kuehl R, Al-Bataineh S, Gordon O, Luginbuehl R, Otto M, Textor M, Landmann R (2009) Furanone at subinhibitory concentrations enhances staphylococcal biofilm formation by luxS repression. Antimicrob Agents Chemother 53:4159–4166

Kumar N, Iskander G (2007) Novel lactams. PCT, patent number WO2007085042

Kumar N, Read RW (2004) Association of antimicrobial compounds with surfaces and polymers. PCT, patent number WO2005053684

Lonn-Stensrud J, Petersen FC, Benneche T, Scheie AA (2007) Synthetic bromated furanone inhibits autoinducer-2-mediated communication and biofilm formation in oral streptococci. Oral Microbiol Immunol 22:340–346

Lonn-Stensrud J, Landin MA, Benneche T, Petersen FC, Scheie AA (2009) Furanones, potential agents for preventing Staphylococcus epidermidis biofilm infections? J Antimicrob Chemother 63:309–316

Lowery CA, Abe T, Park J, Eubanks LM, Sawada D, Kaufmann GF, Janda KD (2009) Revisiting AI-2 quorum sensing inhibitors: direct comparison of alkyl-DPD analogues and a natural product fimbrolide. J Am Chem Soc 131:15584–15585

Ma S, Xie H (2005) Study on halolactamization-gamma-hydroxylation or haloiminolactonization of 2,3-alkadienamides. Tetrahedron 61:251–258

Manefield M, de Nys R, Kumar N, Read R, Givskov M, Peter S, Kjelleberg S (1999) Evidence that halogenated furanones from *Delisea pulchra* inhibit acylated homoserine lactone (AHL)-mediated gene expression by displacing the AHL signal from its receptor protein. Microbiology 145:283–291

Manefield M, Harris L, Rice SA, de Nys R, Kjelleberg S (2000) Inhibition of luminescence and virulence in the black tiger prawn (Penaeus monodon) pathogen Vibrio harveyi by intercellular signal antagonists. Appl Environ Microbiol 66:2079–2084

Manefield M, Welch M, Givskov M, Salmond GP, Kjelleberg S (2001) Halogenated furanones from the red alga, Delisea pulchra, inhibit carbapenem antibiotic synthesis and exoenzyme virulence factor production in the phytopathogen Erwinia carotovora. FEMS Microbiol Lett 205:131–138

Manefield M, Rasmussen TB, Henzter M, Andersen JB, Steinberg P, Kjelleberg S, Givskov M (2002) Halogenated furanones inhibit quorum sensing through accelerated LuxR turnover. Microbiology 148:1119–1127

Manny AJ, Kjelleberg S, Kumar N, deNys R, Read RW, Steinberg P (1997) Reinvestigation of the sulfuric acid-catalysed cyclisation of brominated 2-alkyllevulinic acids to 3-alkyl-5-methylene-2(5H)-furanones. Tetrahedron 53:15813–15826

McCombs JD, Blunt JW, Chambers MV, Munro MHG, Robinson WT (1988) Novel 2(5H)-furanones from the red marine alga Delisea elegans (Lamouroux). Tetrahedron 44:1489–1502

McConnell OJ, Fenical W (1980) Halogen chemistry of the red alga Bonnemaisonia. Phytochemistry 19:233–247

Nasser W, Reverchon S (2007) New insights into the regulatory mechanisms of the LuxR family of quorum sensing regulators. Anal Bioanal Chem 387:381–390

Newman DJ, Cragg GM (2004) Marine natural products and related compounds in clinical and advanced preclinical trials. J Nat Prod 67:1216–1238

Ni N, Choudhary G, Li M, Wang B (2008) Pyrogallol and its analogs can antagonize bacterial quorum sensing in Vibrio harveyi. Bioorg Med Chem Lett 18:1567–1572

Ohta K (1977) Antimicrobial compounds in the marine red alga Beckerella subcostatum. Agric Biol Chem 41:2105–2106

Otter JA, Vickery K, Walker JT, deLancey Pulcini E, Stoodley P, Goldenberg SD, Salkeld JAG, Chewins J, Yezli S, Edgeworth JD (2015) Surface-attached cells, biofilms and biocide susceptibility: implications for hospital cleaning and disinfection. J Hosp Infect 89:16–27

Pawlik JR (1993) Marine invertebrate chemical defenses. Chem Rev 93:1911–1922

Pettus JA, Wing RM, Sims JJ (1977) Marine natural-products XII Isolation of a family of multihalogenated gamma-methylene lactones from red seaweed Delisea-Fimbriata. Tetrahedron Lett 18:41–44

Plavsic M (2004) Long-term variations of folic acid concentrations in the northern Adriatic. Environ Int 30:761–767

Rasmussen TB, Givskov M (2006) Quorum-sensing inhibitors as anti-pathogenic drugs. Int J Med Microbiol 296:149–161

Rasmussen TB, Manefield M, Andersen JB, Eberl L, Anthoni U, Christophersen C, Steinberg P, Kjelleberg S, Givskov M (2000) How Delisea pulchra furanones affect quorum sensing and swarming motility in Serratia liquefaciens MG1. Microbiology 146(Pt 12):3237–3244

Read RW, Kumar N (1999) Production of furanones. PCT, patent number WO1999054323

Ren D, Sims JJ, Wood TK (2001) Inhibition of biofilm formation and swarming of *Escherichia coli* by (5Z)-4-bromo-5-(bromomethylene)-3-butyl-2(5H)-furanone. Environ Microbiol 3:731–736

Ren D, Sims JJ, Wood TK (2002) Inhibition of biofilm formation and swarming of *Bacillus subtilis* by (5Z)-4-bromo-5-(bromomethylene)-3-butyl-2(5H)-furanone. Lett Appl Microbiol 34:293–299

Ren D, Bedzyk LA, Ye RW, Thomas SM, Wood TK (2004a) Differential gene expression shows natural brominated furanones interfere with the autoinducer-2 bacterial signaling system of *Escherichia coli*. Biotechnol Bioeng 88:630–642

Ren D, Bedzyk LA, Setlow P, England DF, Kjelleberg S, Thomas SM, Ye RW, Wood TK (2004b) Differential gene expression to investigate the effect of (5Z)-4-bromo- 5-(bromomethylene)-3-butyl-2(5H)-furanone on *Bacillus subtilis*. Appl Environ Microbiol 70:4941–4949

Shobharani P, Agarwal R (2010) Interception of quorum sensing signal molecule by furanone to enhance shelf life of fermented milk. Food Control 21:61–69

Vervoort HC, Richards-Gross SE, Fenical W, Lee AY, Clardy J (1997) Didemnimides A–D: novel, predator-deterrent alkaloids from the Caribbean mangrove ascidian didemnum conchyliatum. J Org Chem 62:1486–1490

Williams P, Winzer K, Chan WC, Cámara M (2007) Look who's talking: communication and quorum sensing in the bacterial world. Philos Trans R Soc Lond B Biol Sci 362:1119–1134

Yates EA, Philipp B, Buckley C, Atkinson S, Chhabra SR, Sockett RE, Goldner M, Dessaux Y, Camara M, Smith H, Williams P (2002) *N*-acylhomoserine lactones undergo lactonolysis in a pH-, temperature-, and acyl chain length-dependent manner during growth of *Yersinia pseudotuberculosis* and *Pseudomonas aeruginosa*. Infect Immun 70:5635–5646

Zhang R, Chan D, Jessica S, Iskander G, Black DS, Kumar N. (2009) Synthesis of new aryl substituted 5-alkylidenefuran-2(5H)-ones, *ARKIVOC v*, pp 102–115

Zhang R, Iskander G, da Silva P, Chan D, Vignevich V, Nguyen V, Bhadbhade MM, Black DS, Kumar N (2011) Synthesis of new aryl substituted furan-2(5H)-ones using the Suzuki–Miyaura reaction. Tetrahedron 67:3010–3016

Zhu H, Kumar A, Ozkan J, Bandara R, Ding A, Perera I, Steinberg P, Kumar N, Lao W, Griesser SS, Britcher L, Griesser HJ, Willcox MD (2008) Fimbrolide-coated antimicrobial lenses: their in vitro and in vivo effects. Optom Vis Sci 85:292–300

Index

E.P. Ivanova, R.J. Crawford (eds.), *Antibacterial Surfaces*,
DOI 10.1007/978-3-319-18594-1

MIX
Papier aus verantwortungsvollen Quellen
Paper from responsible sources
FSC® C105338
FSC
www.fsc.org

If you have any concerns about our products,
you can contact us on
ProductSafety@springernature.com

In case Publisher is established outside the EU,
the EU authorized representative is:
Springer Nature Customer Service Center GmbH
Europaplatz 3, 69115 Heidelberg, Germany

Printed by Libri Plureos GmbH
in Hamburg, Germany